Springer Theses

Recognizing Outstanding Ph.D. Research

Aims and Scope

The series "Springer Theses" brings together a selection of the very best Ph.D. theses from around the world and across the physical sciences. Nominated and endorsed by two recognized specialists, each published volume has been selected for its scientific excellence and the high impact of its contents for the pertinent field of research. For greater accessibility to non-specialists, the published versions include an extended introduction, as well as a foreword by the student's supervisor explaining the special relevance of the work for the field. As a whole, the series will provide a valuable resource both for newcomers to the research fields described, and for other scientists seeking detailed background information on special questions. Finally, it provides an accredited documentation of the valuable contributions made by today's younger generation of scientists.

Theses are accepted into the series by invited nomination only and must fulfill all of the following criteria

- They must be written in good English.
- The topic should fall within the confines of Chemistry, Physics, Earth Sciences, Engineering and related interdisciplinary fields such as Materials, Nanoscience, Chemical Engineering, Complex Systems and Biophysics.
- The work reported in the thesis must represent a significant scientific advance.
- If the thesis includes previously published material, permission to reproduce this must be gained from the respective copyright holder.
- They must have been examined and passed during the 12 months prior to nomination.
- Each thesis should include a foreword by the supervisor outlining the significance of its content.
- The theses should have a clearly defined structure including an introduction accessible to scientists not expert in that particular field.

More information about this series at http://www.springer.com/series/8790

Kieran L. Hudson

Carbohydrate-Based Interactions at the Molecular and the Cellular Level

Doctoral Thesis accepted by
the University of Bristol, Bristol, UK

Springer

Author
Dr. Kieran L. Hudson
Department of Chemistry
University of British Columbia
Vancouver, BC
Canada

Supervisors
Prof. Derek N. Woolfson
University of Bristol
Bristol
UK

Prof. Timothy C. Gallagher
University of Bristol
Bristol
UK

ISSN 2190-5053 ISSN 2190-5061 (electronic)
Springer Theses
ISBN 978-3-030-08517-9 ISBN 978-3-319-77706-1 (eBook)
https://doi.org/10.1007/978-3-319-77706-1

Printed on acid-free paper

This Springer imprint is published by the registered company Springer International Publishing AG part of Springer Nature
The registered company address is: Gewerbestrasse 11, 6330 Cham, Switzerland

For Dad

Supervisors' Foreword

Kieran Hudson completed this Ph.D. thesis under our joint supervision. This is in the field of carbohydrate–protein interactions, which are fundamental to many biological processes, including: communications between healthy cells and within healthy tissues; the transport of viruses into host cells (infection); and the mechanisms by which pathogenic bacteria colonize niches and avoid detection by the immune system. However, the precise roles played by the carbohydrate (sugar) moieties in these processes remain poorly understood and are challenging to disentangle because carbohydrate–protein interactions are intrinsically weak. Kieran's work addressed issues in this field in two different ways, and, as outlined in this thesis, he has made significant contributions to the area.

The first part of Kieran's thesis describes the first detailed bioinformatics analysis of carbohydrate–protein interactions across the entire database of protein structures, the PDB. Not only does this study furnish us with annotated sets of specific carbohydrate–protein interactions for others to inspect and learn from, but also it uncovers how proteins recognize and discriminate between different sugars. Moreover, Kieran explains his findings in terms of physical chemistry and specifically CH–π interactions associated with the electrostatic properties of the interacting partners. This is distinct from the classical hydrophobic effect and challenges the prevailing view of what stabilizes carbohydrate–protein interactions. Kieran's work was published in the *Journal of the American Chemical Society* in 2015. Further, Kieran was able subsequently to demonstrate that these weak non-covalent CH–π interactions play important roles more generally in stabilizing and specifying protein structures. This work contributed to publication of a further article in *Nature Chemical Biology* in 2017.

The second part of Kieran's thesis put his considerable skills and abilities as a practical synthetic chemist to use. This work is in the area of tissue engineering and the generation of artificial tissues with clear medical applications. For this, soft materials are needed as scaffolds on which cells and tissues can grow. However, to be effective, these materials need to contain chemical and biological cues to promote cell adhesion and growth. Carbohydrates are critical for tissue growth and development in biology, but their complexity means they are under-exploited in the materials science of tissue engineering. Therefore, Kieran developed chemical and enzymatic methods to append sugars to protein-based materials to prepare glycosylated hydrogels, which he then tested for tissue engineering using cell biology. Through this work, Kieran contributed to papers in *ACS Biomaterials Science and Engineering* and *Advanced Healthcare Materials*.

In summary, Kieran has delivered an extremely impressive thesis that combining informatics, physical chemistry and theory, synthetic and biological chemistry, and cell biology.

Bristol, UK
December 2017

Prof. Derek N. Woolfson
Prof. Timothy C. Gallagher

Abstract

Many of the roles of carbohydrates in biology derive from their interaction with proteins, through which they effect intra- and intercellular signalling and the modulation of the structure and activity of proteins. Understanding protein–carbohydrate interactions in atomistic detail is essential to allow the manipulation and exploitation of these processes.

The first part of this thesis utilizes the many published protein X-ray crystal structures that contain carbohydrates. A database of protein–carbohydrate interactions was generated to elucidate the nature of carbohydrate-based interactions at the atomistic and molecular levels. The particular focus is on the carbohydrate–aromatic interaction, involving the positioning of aromatic amino-acid side chains over carbohydrate C–H bonds. This reveals an important role for electrostatics in carbohydrate–aromatic interactions and in CH–π interactions in general, which is distinct from the hydrophobic effect. These findings are supported by solution-phase studies of carbohydrate–aromatic interactions by NMR spectroscopy.

The second part describes the development of carbohydrates as tools for tissue engineering, given the recognized importance of carbohydrates in both signalling and structural roles in biology. The nature of the scaffold upon which artificial tissues are grown is of great importance, as the cellular environment influences development through its physical properties and the presence of biological cues. Carbohydrates are a promising and largely under-exploited class of biomolecules with the potential to modulate material properties and stimulate biological responses.

A modifiable derivative of a system based upon complementary synthetic peptides that self-assemble into hydrogels was used as the core scaffold. This was functionalizable with biological cues *via* copper ‘click’ chemistry. Alkynyl monosaccharides were synthesised and used to verify the applicability of carbohydrates as modifiers, both in terms of maintaining the key properties of the hydrogel and providing an appropriate support for cell culture. Enzymatic techniques enabled synthesis of complex alkyne-functionalised oligosaccharides chosen to be applicable to neural tissue engineering.

Parts of this thesis have been published in the following journal articles:

1. Mehrban, N.; Abelardo, E. S.; Wasmuth, A.; Hudson, K. L.; Mullen, L. M.; Thomson, A. R.; Birchall, M. A.; Woolfson, D. N. *Adv. Healthc. Mater.* **2014**, *3* (9), 1387.
2. Hudson, K. L.; Bartlett, G. J.; Diehl, R. C.; Agirre, J.; Gallagher, T.; Kiessling, L. L.; Woolfson, D. N. J. *Am. Chem. Soc.* **2015**, *137* (48), 15152.
3. Mehrban, N.; Zhu, B.; Tamagnini, F.; Young, F. I.; Wasmuth, A.; Hudson, K. L.; Thomson, A. R.; Birchall, M. A.; Randall, A. D.; Song, B.; Woolfson, D. N. *ACS Biomater. Sci. Eng.* **2015**, *1* (6), 431.

Acknowledgements

First and foremost, I would like to thank my supervisors, Dek and Tim, for their advice and direction over the past four years and for the freedom to pursue research that I found the most interesting and rewarding. They have been constants through the ups and downs, and occasional sideways lurches, inherent in a Ph.D., always on hand to offer guidance and support and this has been invaluable.

As the projects I have worked on have opened up the possibility to try my hand at many different disciplines and techniques, I owe thanks to a large number of individuals for their time and expertise.

I am grateful to everyone who has been part of the Woolfson group during my time here for being welcoming and great to work with. In particular, I would like to thank: Gail Bartlett for being my mentor in coding and bioinformatics and for saving me huge amounts of time with code of her own at the outset of the protein–carbohydrate interactions project; Natasha Burgess for help with electron microscopy; Antony Burton for AUC analysis and for going through the DTC/Ph.D. process with me, if always a little ahead; Nazia Mehrban for helping me with getting started on the hydrogels project, cell culture, and microscopy; Franziska Thomas and Drew Thomson for scientific advice and guidance in peptide synthesis and pretty much everything else; and Alexandra Wasmuth for working with me on rheology.

I have also greatly enjoyed being (one of the last members) in the Gallagher group. I would like to thank: Ian Hazeldon who helped advance the work on the nature of carbohydrate linkers; Jo Sampson for being a friend and support in the laboratory; Charlotte Smith for advice and training in aspects of chemical synthesis; and Nina Ursinyova for timely advice (both chemical and otherwise).

I also owe much to Laura Kiessling, of the University of Wisconsin–Madison, for the inception of the protein–carbohydrate interaction side of my work, and for continuing to collaborate on the project with suggestions and ideas. I would also like to thank her for allowing me to visit her laboratory for a two-month research placement, and her group for being accommodating while I was there, particularly Rob Brown and Roger Diehl, who worked with me on the NMR experiments.

I would also like to thank Steve Withers at UBC in Vancouver, again for accommodating me during a month-long research placement. It is thanks to Steve's generosity and experience, and that of his group, that I have been able to achieve so much in carbohydrate synthesis. I would particularly like to thank Michael Fischer for help with the enzymatic couplings and David Kwan for help with protein expression.

I am indebted to Jon Agirre of the University of York for collaborating on the validation of carbohydrate structures in the PDB, and Gideon Davies who put us in touch. I also owe thanks to Amit Basu of Brown University and Ronald Okoth from his group for providing me with a carbohydrate to use for tissue engineering experiments.

I thank all those involved in the Bristol Chemical Synthesis CDT for funding and for giving me the opportunity to find the project which best suited me and also for valuable training in scientific research.

In addition to funding from the EPSRC and the University of Bristol for my Ph.D. studentship, I have been fortunate to secure funding through several different agencies to further my research and my academic career. I am grateful to: the World Universities Network for funding my placement in Madsion; the RSC for funding my placement in Vancouver and my attendance of the CSC conference while there; the Biochemical Society for contributing to my attendance of the Carbohydrates GRC; and the University of Bristol Alumni Foundation for contributing to my attendance of the International Carbohydrate Symposium in Bangalore.

I have also been supported by the excellent facilities available at the University of Bristol, especially the EMU, MS, and NMR facilities of the School of Chemistry and the Wolfson Bioimaging Facility.

Contents

Abbreviations and Acronyms

anhydr.	Anhydrous
aq.	Aqueous
AU	Absorbance units
AUC	Analytical ultracentrifugation
BCA	Bicinchoninic acid
BME	Basement membrane extract
BSA	Bovine serum albumin
cat.	Catalytic
CD	Circular dichroism
CMP	Cytidine monophosphate
CSS	Carbohydrate Structure Suite
CuAAC	Cu-catalysed azide-alkyne [3 + 2] cycloaddition
DFT	Density functional theory
DIC	Diisopropylcarbodiimide
DIEA	*N,N*-diisopropylethylamine
DMF	*N,N*-dimethylformamide
DMSO	Dimethylsulfoxide
DTT	Dithiothreitol
ECM	Extracellular matrix
EDTA	Ethylenediaminetetraacetic acid
equiv.	Equivalents
ESP	Electrostatic surface potential
FCC	Flash column chromatography
Fmoc	Fluorenylmethyloxycarbonyl
FPLC	Fast protein liquid chromatography
G'	Storage (elastic) modulus
G''	Loss (viscous) modulus
GDP	Guanosine diphosphate
HBTU	*O*-benzotriazole-*N,N,N',N'*-tetramethyluronium hexafluorophosphate

HEPES	4-(2-hydroxyethyl)-1-piperazineethanesulfonic acid
HF	Hartree–Fock
HOBt	1-hydroxybenzotriazole
HPLC	High-performance liquid chromatography
hSAF	Hydrogelating self-assembling fibre
IPTG	Isopropyl β-D-1-thiogalactopyranoside
IR	Infrared
LINUCS	Linear Notation for Unique Description of Carbohydrate Sequences
MALDI	Matrix-assisted laser desorption ionization
MOPS	3-(*N*-morpholino)propanesulfonic acid
MP	Melting point
MRE	Mean residual ellipticity
MS	Mass spectrometry
MTT	3-(4,5-dimethylthiazol-2-yl)-2,5-diphenyltetrazolium bromide
NaPi	Sodium phosphate
NMP	*N*-methylpyrrolidone
NMR	Nuclear magnetic resonance
NSC	Neural stem cell
OPC	Oligodendrocyte progenitor cell
PBS	Phosphate-buffered saline
PCI	Protein–carbohydrate interaction
PDB	Protein Data Bank
Ppb	Parts per billion
RSCC	Real space correlation coefficient
RT	Room temperature
SAF	Self-assembling fibre
SEM	Scanning electron microscopy
SPPS	Solid-phase peptide synthesis
TCP	Tissue culture plastic
TEM	Transmission electron microscopy
TFA	Trifluoroacetic acid
THF	Tetrahydrofuran
TIPS	Triisopropylsilane
TLC	Thin-layer chromatography
TMS	Trimethylsilyl
TOF	Time-of-flight
Tris	Tris(hydroxymethyl)aminomethane
UDP	Uridine diphosphate
UV	Ultraviolet

Standard abbreviations are used for amino acids (three- and one-letter codes) with the addition of azidonorleucine (Anl), which is given the letter code Z, and refer to the L-configuration. Saccharides are referred to using standard notation or according to Linear Notation for Unique Description of Carbohydrate Sequences (LINUCS).[1]

Alkynyl glycosides are referred to with labels. Xxx-n-alkyne refers to the following structure, with the glycoside Xxx.

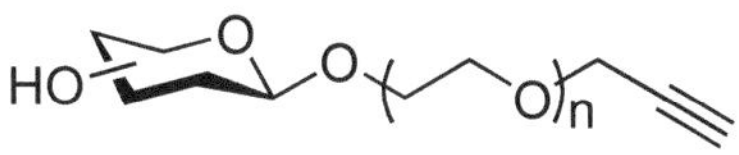

Xxx-n-Ar-alkyne refers to the following structure, with the glycoside Xxx.

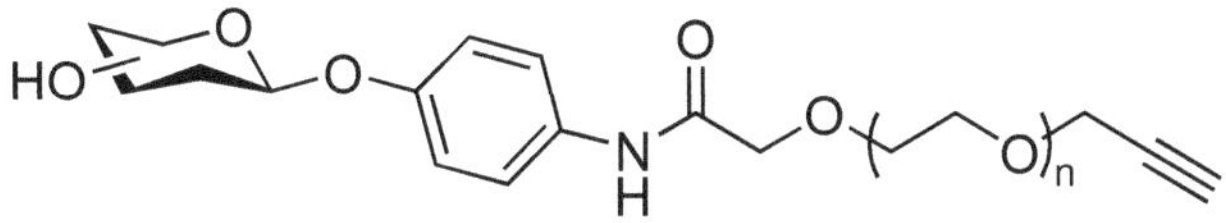

[1]See Bohne-Lang, A.; Lang, E.; Förster T.; von der Lieth, C.-W. *Carbohydr. Res.* **2001**, 336, 1–11.

List of Figures

List of Tables

List of Schemes

Chapter 1
Introduction

1.1 Carbohydrates in Nature

Carbohydrates are ubiquitous in nature, being one of the classes of biomacromolecule (along with proteins, lipids, and polynucleotides) upon which life is based. They are attributed an ever-increasing range of biological roles, due in part to recent progress in glycomics, the identification and study of carbohydrate structure and functions. Glycoscience has been identified as a key area for biomedical research [1, 2].

Carbohydrates perform important structural functions as extended polysaccharides, and also roles in signalling as smaller, but often complex, oligosaccharide glycans [3–5]. These glycans are appended to proteins or lipids to form glycoconjugates and regulate biological processes through interactions with other molecular entities, especially proteins. Specific protein–carbohydrate interactions (PCIs) are implicated in many processes, including human development, cancer, and infectious diseases [6].

Post-translational modification, the appendage of various motifs onto the peptide backbone after protein expression and processing, accounts for much of the function of proteins [7]. Glycosylation is the most frequent and most varied post-translational modification [8], and multiple glycoforms of many proteins are known to exist. These different forms can have completely different functions [6], possibly explaining how organisms achieve complexity from a relatively limited genome.

Many of the functions of glycans are realised directly through intermolecular PCIs, known to be important for processes such as intercellular signalling, infection, and tagging misfolded proteins for degradation [9]. Glycosylation also plays an important role in the structure of glycoproteins through intramolecular PCIs [10]. Glycans can stabilise the correctly folded form of a protein, or alter the conformation adopted. Both this structural role and intermolecular processes rely upon interaction between the specific carbohydrate structure and the amino acids making up the binding protein.

Carbohydrates are also important as constituents of polysaccharides, which feature in the extracellular space of tissues as glycosaminoglycans [11]. When linked to extracellular proteins to form proteoglycans these are involved in binding

K. L. Hudson, *Carbohydrate-Based Interactions at the Molecular and the Cellular Level*, Springer Theses, https://doi.org/10.1007/978-3-319-77706-1_1

protein ligands intra- and intercellularly and altering the mechanical properties of the extracellular matrix (ECM). As well as physical roles, ECM polysaccharides such as hyaluronan have been implicated in cellular behaviour [12].

1.1.1 Carbohydrate Structure

Carbohydrates can achieve huge diversity of structure. This is because the monosaccharide monomers can be linked to each other through any one of several chemically similar hydroxyl motifs. Monosaccharides can also exist in the D- or L-configuration, in several ring forms, and as either an α- or β-anomer. The type and position of linkages between monosaccharides defines oligosaccharide structure and therefore function.

Pyranose and Furanose Forms and D- and L-Configuration

A carbohydrate consists of an aldehyde (or ketone) bearing a chain of carbons substituted with hydroxyl groups (**1**, Fig. 1.1). The length and stereochemistry of this chain varies between monosaccharides. The carbohydrate carbons are numbered systematically from the aldehyde as C1, with substituents following this numbering, e.g., the atoms bound to C2 are labelled H2 and O2.

The same monosaccharide can exist in a number of structural forms (Fig. 1.1). The acyclic form (**1**) is not commonly seen, as cyclisation by reaction of a hydroxyl with the aldehyde to give a hemiacetal is favoured. This typically takes place through the O4, to give the five-membered furanose form (**2**), or through O5, to give the six-membered pyranose form (**3**). These two ring forms can exchange in aqueous solution, and so the lowest energy ring is favoured and seen the most often in nature. Whether the furanose or pyranose form is favoured differs between monosaccharides.

Monosaccharides can also exist in D- or L-forms, determined by the configuration around the stereogenic carbon farthest from the aldehyde/ketone (C5 for a hexose, **1**), which are enantiomers. For most monosaccharides the D-form (**3**) is more common in nature, but in a few it is the L-form, e.g., L-fucose (Fuc). Six-membered rings, and

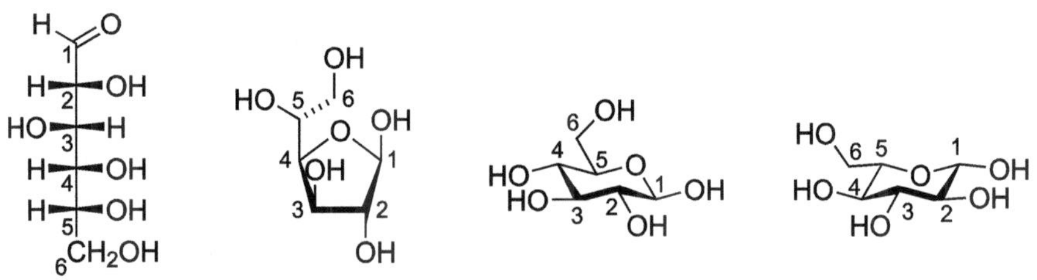

1 **2** **3** $^4C_1$ **4** $^1C_4$

Fig. 1.1 Four forms of glucose, with carbons numbered systematically: acyclic (**1**), furanose (**2**), and pyranose (**3**) forms of D-glucose (Glc), and D- (**3**) and L-configurations (**4**) of glucopyranoside

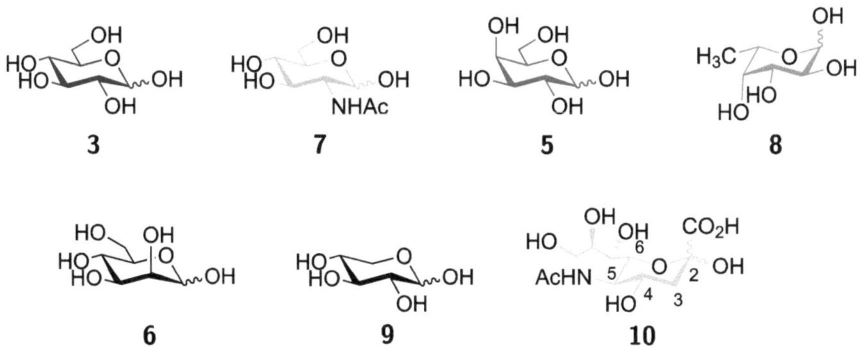

Fig. 1.2 Structures of some of the commonly occurring natural monosaccharides featured in this work: Glc (**3**), GlcNAc (**7**), Gal (**5**), Fuc (**8**), Man (**6**), Xyl (**9**), and NeuNAc (**10**). Colour codes are used throughout this thesis

hence pyranoses, can adopt a number of conformations, of which the lowest energy is normally a chair conformation, either $^4C_1$ for the D-form (C4 above C1, **3**) or $^1C_4$ for the L-form (C1 above C4, **4**) [13].

Structures of Common Monosaccharides

The work presented in this thesis addresses and utilises mammalian monosaccharides (Fig. 1.2), as I primarily targeted human therapy. All of these favour the pyranose form, and are always referred to in their most common configuration (all D except for Fuc). Several of these are simple hexoses, including Glc (3), D-galactose (Gal) (**5**), and D-mannose (Man) (**6**). These only differ by the configuration of certain hydroxyls, and yet fulfil very different roles in biology. The 2-*N*-acetylamino derivatives of several monosaccharides also often occur in nature, such as N-acetly-D-glucosamine (GlcNAc) (**7**). Fuc (**8**) has a methyl group at C6, and is the equivalent of 6-deoxy-L-galactose. D-xylose (Xyl) (**9**) is a pentose, being the equivalent of Glc without the C6 hydroxymethyl. *N*-acetylneuramic acid (NeuNAc) (**10**) is a more complex monosaccharide, which still favours a six-membered pyranose form, with a carboxylic acid at the anomeric C2, no substituent at C3, and a three carbon glycol chain at C6.

Anomers and Glycosidic Linkages

The cyclisation of acyclic monosaccharides through hemiacetal formation generates a new stereocentre at the C1 carbon. The two possible cyclic forms are known as anomers—in pyranoses the β-anomer has the anomeric substituent *cis* (on the same side of the ring) to the C5 substituent, and for the α-anomer the anomeric substituent is *trans* to the C5 substituent (Fig. 1.3).

In oligosaccharides, monosaccharides are most commonly linked to others at the anomeric centre. These glycosidic linkages between monosaccharides can be either α or β, depending on the anomeric form of the C1-linked unit (Fig. 1.3). The second monosaccharide can be linked through any hydroxyl, and which is involved is denoted in the linkage type, e.g., β-1,4-glucopyranosyl β-glucopyranoside (**13**)

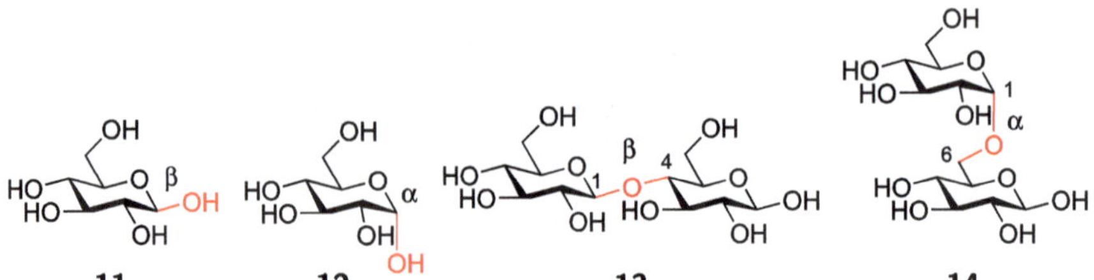

Fig. 1.3 Glc as β- (**11**) and α-anomers (**12**), and linked in β-1,4 (**13**) and α-1,6 (**14**) glycosidic linkages. Anomeric groups and linkages highlighted in red

involving O4 or α-1,6-glucopyranosyl β-glucopyranoside (**14**) involving O6. Single monosaccharides can be linked to more than two others in oligosaccharide chains, leading to branching and greatly extending the number of possible structures that can be generated from relatively few units.

LINUCS [14] is a notation system allowing these complex carbohydrate structures to be written in a single line and in a form that can be interpreted by computer programmes. Square brackets [] denote monosaccharide (standard labels with an additional p for pyranose or f for furanose), amino acid, or other units; braces { } are used to enclose branches; parentheses () show the linkages, with a and b for α and β, respectively. For example, structure β-1,4-glucopyranosyl β-glucopyranoside (**13**) would be denoted [][b-D-Glcp]{[(4+1)][b-D-Glcp]{}}. This notation will be used in the discussion of the analysis of PCIs (Part I).

1.2 Protein–Carbohydrate Interactions

1.2.1 The Roles and Importance of Protein–Carbohydrate Interactions in Nature

Many of the roles that carbohydrates play in nature are achieved through interaction with proteins [3, 4, 9, 15], through intermolecular interactions (as ligands or glycoconjugates) and intramolecular effects on protein structure (as glycans) [16]. Proteins that bind carbohydrates non-covalently are known as lectins, and these are implicated in many aspects of biology, including intercellular and cell–matrix interactions and inflammation [9, 15]. All of these roles depend on the recognition of specific oligosaccharide units by the proteins concerned. The activity of lectins depends on binding the correct glycan appended to the glycoconjugate being bound. This means that carbohydrate-processing enzymes, such as glycosyltransferases and glycosidases, that build these glycans are equally as important as lectins.

PCIs are valuable targets for therapy as this class of interaction is implicated in many aspects of disease [17, 18]. Infectious agents use carbohydrate epitopes displayed on the cell surface for recognition and entry [17], such as the influenza virus that binds to cell-surface sialic acids, and the disruption of these recognition processes could lead to prevention of disease. Unique glycans not found in the human body are presented on the surface of many pathogens, making promising targets as the bases for vaccines [19]. Abnormal glycosylation is also a hallmark of cancer cells, and so targeting PCIs represents a potential route to treatment, and the interaction of the aberrant glycocalyx with proteins contributes to cell growth and survival [20].

PCIs also have biological effects through modulation of the structure and properties of glycoproteins [10]. The effects of glycosylation can be general, such as masking hydrophobic patches on a protein surface to prevent aggregation or enhancing glycoprotein stability and accelerating folding [21]. The latter can occur through destabilisation of unfolded states by steric exclusion, or formation of more-specific favourable interactions with amino acids [22].

The promiscuity of most carbohydrate-active enzymes means that many glycoproteins exist as more than one glycoform, and glycosylation is of sufficient importance that glycoforms can have completely different effects in vivo. For example, different glycoforms of immunoglobulin G have been shown to have a range of effects, including pathogenicity, pro-, and anti-inflammatory responses [6].

PCIs engaged in by lectins and carbohydrate-active enzymes have developed in nature and undergone extensive optimisation through the process of evolution. Given the variety of chemical functionality available in the proteinogenic amino acids, structures of protein–carbohydrate complexes represent an excellent source of data to further the understanding of carbohydrate-based interactions.

1.2.2 Analysis of Protein–Carbohydrate Interactions

The recognised importance of carbohydrate-based interactions means that many different techniques have been employed in studying carbohydrates [9, 23]. The field of glycomics addresses the identification and analysis of glycan structures, which is often a challenge due to the complexity of structures with very minor chemical differences [23]. Glycomics relies on PCIs, utilising the specific binding of lectins to identify oligosaccharides, and the known modes of action of glycosidases to determine linkages. The interactions of carbohydrates with proteins can also be probed directly, including the molecular basis by computational modelling, solid-state structural analysis, Nuclear Magnetic Resonance (NMR) spectroscopy, or quantifying the energy of interactions through biophysical techniques, affinity analysis, or isothemal titration calorimetry [9].

Computational modelling has become a powerful technique for analysing PCIs, due to substantial improvements in force fields appropriate for use with carbohydrates [24–26]. The computational docking of oligosaccharides into binding sites gives insight into the key interactions that drive association, and also the underlying dynamics. The model can be altered to determine the importance of different factors in binding, for example removing particular hydrogen-bonding amino acids or adjusting the relative contribution of factors, such as hydrophobic effects, to the force field. Altering the strengths of contributions from interactions with polar and non-polar amino acids in a simulated hevein–GlcNAc complex shows that both classes must be removed to entirely abrogate the binding [27].

Several techniques can be used to analyse PCIs in solution, i.e., in the biological state [9]. NMR spectroscopy is one of the most powerful, giving information on the carbohydrate conformation, binding dynamics, and even molecular detail of specific interactions such as hydrogen bonds [28]. These experiments can be carried out on model systems or for carbohydrates interacting with whole proteins, and vary from simple interpretation of ^{1}H chemical shifts to isotopic labelling and two-dimensional techniques. The combination of NMR and computational experiments gives detailed insights into specific PCIs, for example identifying the key interactions in the binding of β-galactosides by galectin-1 as hydrogen bonds to C(4)–OH and C(6)–OH of the Gal residue [29].

Another powerful approach for elucidating PCIs is to modify, or design *de novo*, species that bind carbohydrates [30]. The effect of directed mutagenasis experiments on the carbohydrate binding of lectins or activity of carbohydrate-active enzymes can highlight key factors behind the interaction. The synthesis of artificial lectins based upon binding principles can verify those ideas, and give tools to probe carbohydrate binding in more detail.

X-Ray Protein Crystallography: The Protein Data Bank

X-ray crystallography can be used to obtain the three-dimensional structures of carbohydrates, proteins, and protein–carbohydrate complexes in the solid state. This gives information on PCIs at the atomic level, showing the specific modes of interaction [9]. The RCSB Protein Databank (PDB) is a data bank of protein structures [31], with the majority obtained by X-ray crystallography, and represents an ever-increasing repository of data for the identification of trends in structural biology [32].

As of 2011, 6% of crystal structures in the PDB contained carbohydrates, of which 40% were proteins co-crystallised with carbohydrates and the remainder glycoproteins [33]. The total number is increasing rapidly with the size of the PDB (Fig. 1.4), and the rate of growth increases each year. Therefore, there is much information about the nature of carbohydrate-based interactions that can be inferred from the PCIs captured in the PDB. It is important to note that crystal structures represent the optimal non-localised interactions of entire species—while specific intermolecular bonds contribute to these interactions, the overall context must be considered [34].

The utility of the data held within protein crystal structures for understanding carbohydrate-based interactions has been recognised [36], and as such many computational tools and on-line databases have been designed to aid in the categorisation and interpretation of this data [37]. Worthy of particular mention are the Carbohydrate-Active Enzyme (CAZy) database [38], which focusses on classifying proteins active as enzymes, and the Carbohydrate Structural Suite (CSS) [35], which is concerned

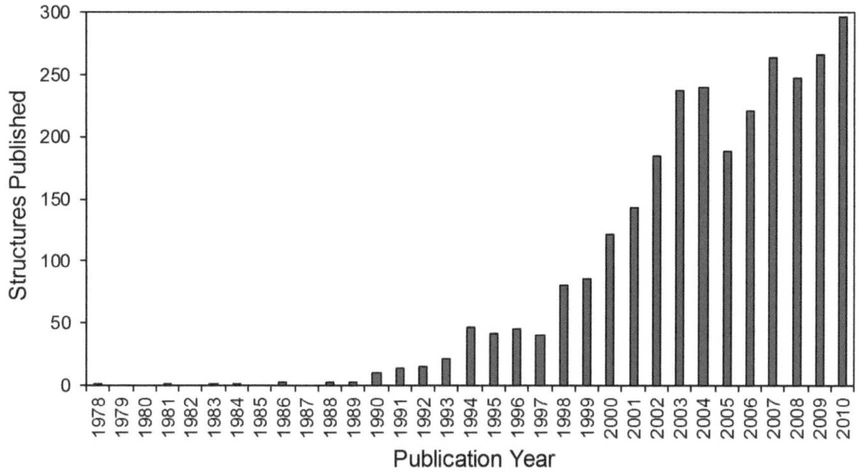

Fig. 1.4 Structures contained within GlyVicinity [35] database of carbohydrate-containing PDB entries by publication year. The last update before the data was obtained occurred in mid-2011

primarily with the carbohydrates themselves. The CSS contains tools to identify carbohydrates within crystal structures in the PDB, classify their conformations, and analyse interacting amino acid residues.

Several computational methods have been developed to use the information about PCIs within carbohydrate-containing protein structures [39–42]. These have all involved the statistical analysis of a training set of carbohydrate-binding proteins, from nineteen [39] to almost 500 [42] examples, for the identification of binding sites in further proteins. Prediction accuracies of 60% to 90% are achieved, and some features of PCIs such as a preference for the aromatic amino acids tryptophan (Trp) and tyrosine (Tyr) in the binding sites are evident. However, focussing on the binding site and not distinguishing between monosaccharides restricts the amount of detail that can be inferred about carbohydrate binding and discrimination.

Other studies look at the amino acid environments of carbohydrates bound to proteins in more detail to derive this information [43, 44]. These have been limited to determining the trends in binding for a single class of monosaccharide [45, 46] or protein [47]. Therefore, they show that there are specific binding patterns for different carbohydrates, but have not had the breadth to establish general rules for how carbohydrates interact. A more detailed study has been carried out of the amino acid environments of glycans appended to glycoproteins [48]. This illustrates an influence of local sequence near to glycosylation sites on the glycan, and preferences for certain residues at points in the sequence relative to the site. Again it is the aromatic amino acids that are particularly over-represented near to glycans.

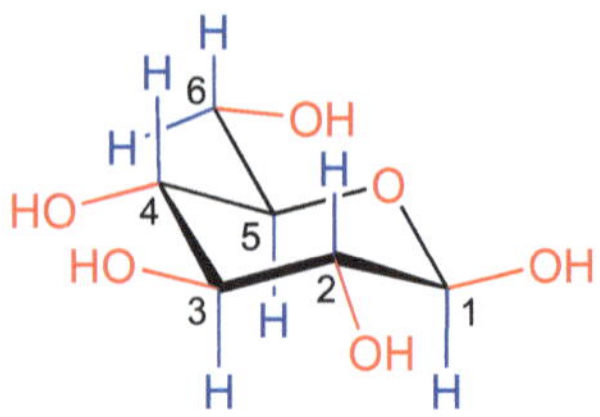

Fig. 1.5 Structure of β-D-Glc, with C–OH bonds highlighted in red and C–H bonds highlighted in blue

1.2.3 The Molecular Basis of Protein–Carbohydrate Interactions

In order to exploit the many important roles played by PCIs in nature, it is necessary to understand the interactions at the molecular level. The many techniques applied to the study of PCIs means that common features of carbohydrate binding have been identified [9]. All carbohydrates have two functional groups in common—hydroxyls and C–H bonds, Fig. 1.5—with some having carboxylate, amine, sulfate, and many other possible functionalities. The structure of a carbohydrate determines the conformation and hence how these groups are presented in three dimensions, and carbohydrate-binding sites are optimised to interact with this presentation [49]. The carbohydrate hydroxyls engage in hydrogen bonding with polar amino acid side chains, and electrostatic interactions involving charged groups are also possible for acidic or amino sugars. Also commonly identified are 'non-polar' interactions, involving the carbohydrate C–H groups and hydrophobic amino acids. For many specific oligosaccharides, the binding sites are very similar across different kingdoms of life, reflecting both conservation through evolution and convergence to common solutions [50].

Given the reliance on hydrogen-bonding and non-polar interactions, the binding energies of individual monosaccharide residues to proteins are often small. Binding at the level required for biological relevance is achieved through multivalent interactions [51, 52], referred to as the 'cluster glycoside' effect for sugars [53]. Binding of multiple motifs attached together is stronger than the sum of the individual interactions due to favourable entropy from an increased effective local concentration; binding one of the motifs brings the others into closer proximity of the receptors, making subsequent binding events more likely (Fig. 1.6). This effect is at both the molecular level, with binding enhanced for oligosaccharides versus monosaccharides, and cellular level in the presentation of many copies of the same glycan on protein and cell surfaces.

The hydroxyl moieties have traditionally been considered to be the most important motifs of carbohydrates, [9, 54–56] as changes in hydroxyl configuration define monosaccharide identity and hence function. Hydroxyls do indeed contribute significantly to binding through the formation of hydrogen bonds (Fig. 1.7), and single hydrogen bonds can be necessary for the maintenance of entire PCIs, e.g., for binding by selectins [57]. This contribution to binding is primarily an entropic effect [58].

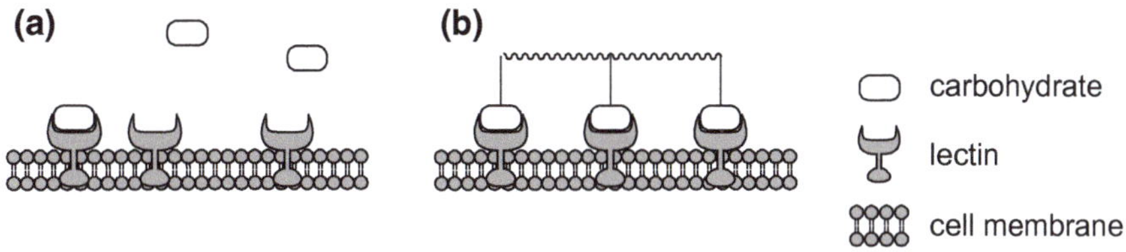

Fig. 1.6 Importance of multivalency in carbohydrate binding. **a** Individual PCIs are weak and transient. **b** Multivalent interactions, with multiple binding epitopes linked together, are enhanced due to avidity (entropy effects and additive binding enthalpy)

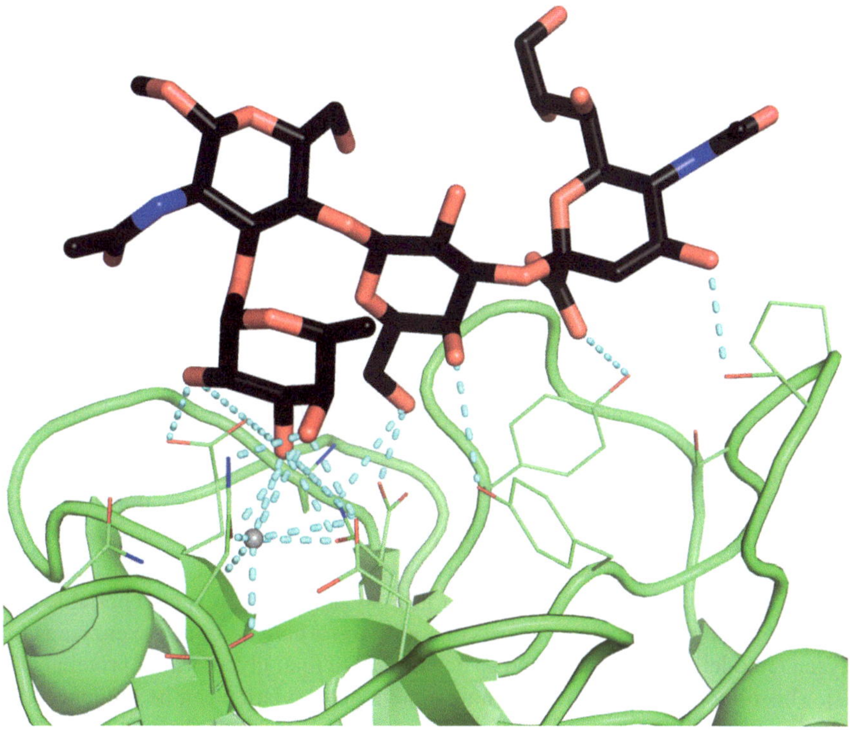

Fig. 1.7 The binding of the oligosaccharide sialyl-Lewis^X (SLe^X), in black, by P-selectin, green, is achieved through multiple hydrogen bonds with the carbohydrate hydroxyls (shown as cyan dashed lines). Divalent cations such as Ca^{2+} (shown as a grey sphere) often act as hydrogen-bond acceptors in binding sites. Image generated from PDB ID 1G1R using PyMOL [59]

Energetically, hydroxyls resemble water molecules, the ubiquitous solvent in nature, and so the enthalpic contribution compared to binding water is negligible.

Weak non-covalent interactions between relatively apolar, hydrophobic regions of species are often key to molecular recognition in water [60]. Carbohydrates have such regions where clusters of C–H bonds are presented, often referred to as 'non-polar patches' [9, 15], determined by the hydroxyl configurations and hence monosaccharide identities. The preference for these areas of carbohydrates to interact with hydrophobic amino acid residues, particularly aromatic ones, in protein binding sites is long established [54].

1.2.4 Carbohydrate–Aromatic Interactions

Given the frequent observation of aromatic residues in carbohydrate binding sites, carbohydrate–aromatic interactions are increasingly the subject of study [61]. In some cases, interactions with arenes have been shown to be key to the binding of carbohydrates, for example in human lysozyme where mutation of a Tyr to a (still hydrophobic) leucine (Leu) residue reduces efficacy of the enzyme [62]. Carbohydrate–aromatic interactions have also been implicated in the transport of carbohydrates through transmembrane proteins [63], pre-organising glycans into active conformations [64], and stabilising the folding of proteins [65]. These interactions have also been identified as being important in the modes of action of certain drugs [66], and in mechanisms of DNA repair [67].

The propensity for carbohydrates to interact with arenes has been exploited in the synthesis of artificial carbohydrate-binding systems [30]. The presentation of C–H bonds on both the α- and β-faces of β-D-Glc and its derivatives can lead to interactions with aromatic residues presented to both faces simultaneously, in a so-called 'aromatic sandwich' arrangement (Fig. 1.8) [68]. This has formed the basis of synthetic lectin design [69].

Recent evidence has suggested new and important roles for intramolecular carbohydrate–aromatic interactions. Interactions with amino acids in the protein can alter how tightly bound glycans are, and hence the dynamics of the oligosaccharide and of the protein chain near to the site of glycosylation. This can influence activity of binding regions near to the site of glycosylation [70]. It is also possible that mobility of the glycan determines accessibility to carbohydrate-active enzymes for glycan remodelling in the cell, and so determine the final glycoform of proteins [71, 72]. This offers an explanation for how protein glycosylation can be influenced by genetics, despite the promiscuity of most carbohydrate-active enzymes.

CH–π Interactions

It is recognised that aromatic residues commonly interact with the C–H groups of carbohydrates. Aromatic amino acids are relatively hydrophobic [73], and it has been suggested that the contribution to binding is through a hydrophobic effect [74]. However, gas-phase experiments discount this as interactions are still observed in the absence of any solvent [75]. Aromatic rings present electronegative π-systems above and below the ring, and these play an important role in aromatic interactions [76, 77]. C–H bonds are polarised, meaning that the proton is partially positive, and have a favourable electrostatic interaction with the aromatic π-systems. These contribute to so-called CH–π interactions (Fig. 1.9) [43, 78].

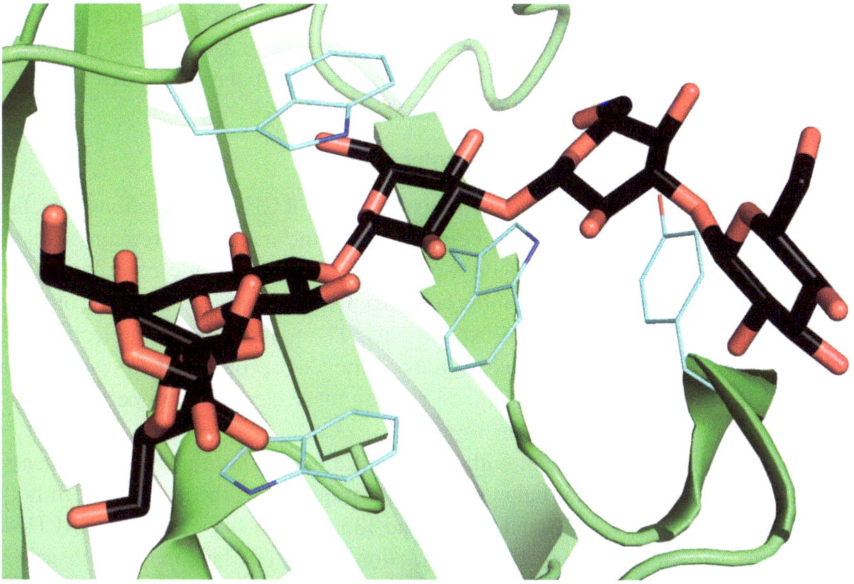

Fig. 1.8 Several aromatic residues (highlighted in cyan), complemented by hydrogen bonds (not shown), are key to the binding of glucose-based saccharides, in black, by carbohydrate-binding module 4, green. Image generated from PDB ID 1GUI using PyMOL [59]

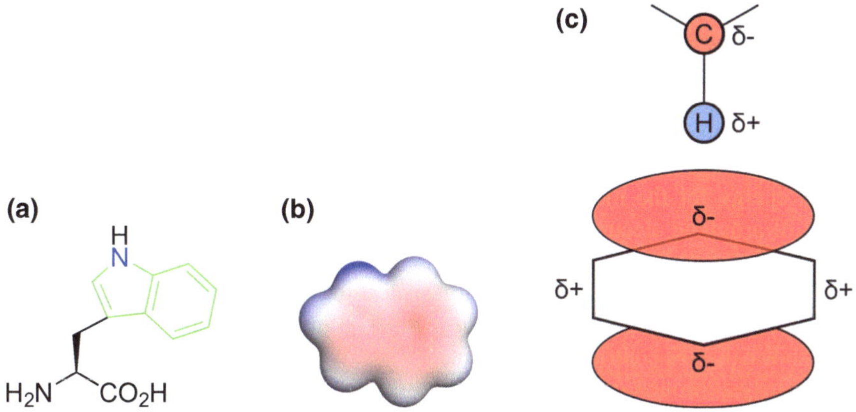

Fig. 1.9 **a** Structure of Trp, with aromatic ring (indole) highlighted in green. **b** Electrostatic surface potential of indole, with negative regions coloured red and positive regions coloured blue. **c** A CH–π interaction is between the partially positive C–H proton and the negative charge of the aromatic quadrupole

Intramolecular CH–π interactions involving aromatic amino-acid side chains are frequently observed in protein X-ray crystal structures [79], and have been detected in solution by NMR spectroscopy by the observation of through-space coupling between the carbohydrate nuclei and aromatic ring [80].

Electrostatic Contribution to CH–π Interactions

Several studies have tried to elucidate the exact nature of CH–π interactions [43], but the relative importance of different components of the interaction remains subject to debate. While it is acknowledged that there is an electrostatic contribution, as well as van der Waals and hydrophobic components, the relative importance of these differs in the findings of different experiments. The magnitude of an electrostatic contribution will depend on the electronics of the C–H proton and π-system. If it is important for CH–π interactions, then varying these electronic properties should have an effect on the interaction energy.

This approach has been taken with several different solution-phase experiments. NMR studies with molecular balances show an influence of the electronics of both the aromatic system [81, 82] and C–H bond [83, 84] on the formation of CH–π interactions. As would be expected, interactions are more favourable with more-electropositive C–H protons and more-electron-rich arenes. Infrared (IR) spectroscopic studies also confirm an influence of C–H electronics on the formation of interactions for a series of substituted phenols [85], although, while electron-withdrawing substituents reduce the interaction, electron-donating groups had little influence.

Computational studies give a more varied picture, showing an important electrostatic contribution only for relatively polarised C–H bonds, such as those in acetylenes [86, 87] (Table 1.1). Interactions of sp-, sp^2-, sp^3-hybridised C–H protons with substituted benzenes have shown a dependence on the aromatic electronics [88, 89]. Indeed, in these calculations the dispersion and exchange energies cancel each other out and so it is the electrostatics that are responsible for the majority of the interaction energy for the more-electropositive sp and sp^2 cases. Again the observed effect of aromatic electronics is greater for electron-withdrawing groups than for electron-donating.

That mesomerically electron-donating substituents do not appear to enhance CH–π interactions might be explained by further computational experiments, which have shown that the contribution of aromatic substituents to non-covalent interactions is primarily a through-space effect, rather than due to changes in the electronics of the π-system itself [90, 91]. The importance of this direct electrostatic interaction of ring substituents has been shown in similar solution-phase experiments investigating aromatic stacking interactions [92].

Similar substituent effects have been observed in experiments investigating cation–π interactions [93]. These interactions, between a positively-charged species and a π-system [94, 95], can be considered to be analogous to CH–π interactions. In the case of organic cations, such as the tetramethylammonium cation, the positive charge resides largely on C–H protons.

Table 1.1 Electrostatic contribution to total calculated energy for computed interactions, selected published data [86, 88]

Interaction		Energy/kcal mol^{-1}		Reference
Donor	Acceptor	Electrostatic	Total	
Hydrogen bond				
H_2O	H_2O	−6.65	−4.80	[88]
Cation-π interaction				
$N(CH_3)_4^+$	C_6H_6	−9.66	−5.77	[88]
CH–π interaction				
CH_4	C_6H_6	−0.19	−1.47	[88]
CH_3F	C_6H_6	−0.93	−2.31	[88]
CH_2F_2	C_6H_6	−1.55	−3.22	[88]
CHF_3	C_6H_6	−2.43	−4.18	[88]
CHCH	C_6H_6	−1.70	−2.75	[88]
		−2.89	−2.45	[86]
CHCH	C_6F_6	−0.23	−0.03	[86]
CHCH	$C_6(CH_3)_6$	−5.08	−4.04	[86]

Forces in Carbohydrate–Aromatic Interactions

It is recognised that CH–π interactions play an important role in carbohydrate–aromatic interactions [61]. Many studies have been conducted to offer insights into the nature of carbohydrate–aromatic interactions, and the components that are important in defining them.

One of the best-developed techniques for investigating CH–π interactions of carbohydrates and carbohydrate-like species is NMR spectroscopy [28]. Participation in a CH–π interaction reduces the ^{1}H NMR δ shift for a C–H proton, due to the ring currents of the aromatic π-system, which lead to an induced magnetic field over the centre of the ring that is opposed to the applied magnetic field [96]. The magnitude of the difference in C–H proton δ (Δδ) for species in the presence compared to the absence of an aromatic species is therefore indicative of the strength of CH–π interactions being formed, ranging up to values of around 10 Hz [97]. Studies with dynamic combinatorial libraries show an influence of the electronics of the carbohydrate C–H bond and of the aromatic system on energetics of carbohydrate–aromatic interactions; pairing of electron-poor C–H bonds with electron-rich aromatics can lead to stability increases in complexes of up to 1.5 kcal mol^{-1} in ΔG [98].

A model system based on a β-hairpin peptide shows the importance specifically of CH–π interactions in carbohydrate–aromatic interactions, as a favourable influence on binding is only observable with aromatic rings and not with cyclohexane, changing from 0 to -1 kcal mol^{-1} ΔG of folding [99]. There are also slightly stronger interactions with more-electron-rich aromatic rings, and for carbohydrates with hydroxyls protected with electron-withdrawing acetyl groups over electron-donating methyl groups. Binding of carbohydrates by synthetic lectins can also be increased approximately 5-fold by using aromatic platforms with electron-donating substituents [100]. In contrast, recent experiments investigating the folding of a different peptide system, a WW-domain, as a model of short-range interaction of GlcNAc in a protein turn shows little dependence on the electronics of the arene [101]. A study on the CH-π interaction of different carbohydrates and derivatives with DNA bases also did not show a correlation with C–H electronics [102].

Computational studies have confirmed the energetically favourable interaction of carbohydrate C–H bonds and aromatic rings, and show that different monosaccharide residues have different preferred modes of interaction [103–105]. The contributions of interactions of carbohydrates with specific arenes in protein binding sites have been calculated, and confirmed by mutagenesis experiments [106]. An interesting observation from calculated interaction energies is that, while the 'bidentate' interaction of two carbohydrate C–H bonds with a single aromatic system is more favourable than an equivalent interaction of a single C–H bond, it is less favourable than two individual interactions [107], suggesting a degree of cooperativity; even in a 'monodentate' system the proximal second C–H proton would make some through-space electrostatic contribution to binding, emphasising that carbohydrate–aromatic interactions derive from the entirety of the two species interacting and not specific intermolecular bonds.

1.3 Scope of the Thesis Part I: Analysis of Protein–Carbohydrate Interactions in Protein Structures

In this study, I have used the PDB to understand PCIs, taking advantage of the recent growth in size of this database to perform an in-depth analysis. I included all classes of protein and carbohydrates in order to gain the most general understanding possible. The statistical analysis of the amino acids surrounding different carbohydrates in protein X-ray crystal structures gives insights into the basis of carbohydrate-based interactions, and specifically how particular carbohydrates are bound and how proteins discriminate between very similar glycans.

Firstly, I developed computer scripts to identify carbohydrate residues and their nearby amino acids in protein structures. This required validation of the data from the PDB and the classification of appropriate species. I also developed techniques to analyse particular carbohydrate-based interactions that are of interest. This will focus on carbohydrate–aromatic interactions in particular, given their recognised but ambiguous role in PCIs. This means looking at how the C–H bonds of carbohydrates interact with different amino acids, and the identification of CH–π interactions.

Secondly, I modeled the properties of the carbohydrates to determine differences between the C–H bonds, and compare this to the identified interactions with aromatics. This determines the importance of electrostatics in carbohydrate–aromatic interactions in proteins, and CH–π interactions more generally.

Finally, I determined how these findings relate to interactions in solution by carrying out NMR spectroscopy studies based upon my findings from the crystallographic database.

Establishing the importance of electrostatics in carbohydrate–aromatic interactions enhances understanding of the molecular basis of PCIs. This enables the design of better drugs to target the many biological processes involving carbohydrates, and also aids with the characterisation of new processes.

1.4 Tissue Engineering

Tissue engineering is an interdisciplinary field devoted to developing artificial tissues to replace or supplant biological functions [108–110]. It combines materials science, biomedical engineering, and cellular biology to understand tissue function and to ultimately design and apply tissues to patients directly, or in combination as entire synthetic organs (Fig. 1.10).

The tenet of tissue engineering is that the use of *de novo* tissues in situations such as wound treatment and organ regeneration or replacement will be beneficial over current synthetic options. This is because the engineered tissues can interact with the existing biological systems as if native, and there is no need for renewal of the material or removal when repair is complete. Progress in the field has accelerated rapidly since its inception in the 1990s, allowing creation and even implantation of organs partially or wholly consisting of artificial tissue into live subjects [112, 113].

Successful tissue engineering requires suitable cells, and then the ability to manipulate and direct them to develop into the desired tissue. In nature, the progenitors of tissues are various classes of stem cells [114], and these are generally the basis of tissue engineering technologies. An understanding of what determines the key behaviours of developing cells, such as adhesion, proliferation, and differentiation, is essential. The nature of these factors can be derived from the biological environments where the tissues develop naturally, and the key properties replicated in the artificial system.

Combinations of multiple environmental factors are known to determine cellular development [115]. These factors include the binding of biological cues, such as growth factors and small molecules [116], as well as the physical properties of the three-dimensional environment in which the cells grow, such as its stiffness [117] and topography [118]. In vivo, these factors are partly defined by the ECM, the milieu of biomolecules that surround the extracellular space [119], and so recreating and controlling the elements of the ECM synthetically has emerged as one of the key challenges in tissue engineering [120].

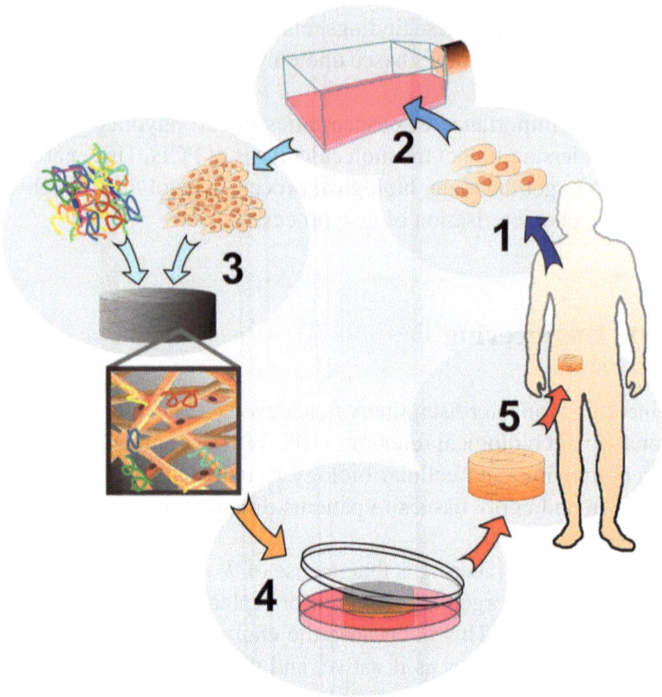

Fig. 1.10 Cartoon representation of the principles of tissue engineering. **1.** Appropriate cells are sourced, ideally from the patient. **2.** Cells are selected and suitable cells proliferated. **3.** The cells are seeded on an appropriate scaffold as a substrate for cell culture. **4.** The seeded cells are cultured to form an artificial tissue. **5.** The generated tissue is implanted into a patient. Figure reproduced from George (2009) with permission of the author [111]

1.4.1 Tissue Engineering Scaffolds

Much research has been focussed on engineering supports for mammalian cell culture, known as scaffolds, with precise control of constituents and properties [121]. These are required to maintain form at biological pH and temperatures, and have mechanical properties resembling the target tissues. This must be achieved while allowing diffusion of H_2O and additives ranging from small molecules to growth factors freely through the system [115]. The ability of the material to persist for several weeks is essential to allow the tissues time to develop, but degradation to harmless by-products once the tissue becomes self-supporting is also beneficial [122]. Minor environmental changes greatly influence cellular behaviour [120], so fine control of each of these characteristics is highly desirable. In biological systems, the local cellular environment, or 'cellular niche', is defined in three dimensions, and even the thinnest tissues are three-dimensional. Indeed, cells behave differently in three-

Table 1.2 Approximate stiffness (elasticity) of various mammalian tissue types [115]

Tissue	Elasticity range/kPa
Brain	0.2–1
Fat	2.5–4
Muscle	8–11
Cartilage	11–12
Bone	12–14

dimensional to two-dimensional environments, and so to properly understand and control cellular development three-dimensional scaffolds are essential [123].

The appropriate range of scaffold stiffness varies for different tissue types, from about 200 Pa elasticity for soft neural tissue to over 10 kPa for bone (Table 1.2) [115]. The stiffness of a scaffold can determine the lineage taken by certain classes of stem cell [124], and this effect is independent of other factors such as protein tethering and scaffold porosity [125].

Hydrogels for Tissue Engineering

Scaffolds for cell culture can be derived from many different materials. These materials are usually hydrogels [126–128], formed from fibrous materials that entangle to a solid mesh with a high water content. Being mostly water makes them ideal for tissue engineering applications, as there is little of the scaffold material to degrade or cause deleterious side effects after the tissue is established. It also allows the required diffusion of nutrients and oxygen to the cells, and of waste materials away, which is essential in three-dimensional cellular environments.

Many hydrogels are based on organic or inorganic synthetic polymers, such as poly(α-esters) and polyurethanes [126, 128]. These have several advantages, mainly a result of their extensive previous use and development in other fields of materials science. This means that, as well as being cheap and readily available in large quantities, they are well understood and their properties can be defined effectively. Hydrogels derived from synthetic polymers are also minimal systems, made up of few chemically inert components. However, this simplicity can mean that they lack scope for adaptation or modification. Also, although they can closely resemble natural supports macroscopically, on the molecular level they differ significantly. This raises the possibility of incompatibility in vivo and, while degradable synthetic polymers have been developed, the degradation by-products are also not natural and potentially toxic.

To overcome these compatibility and toxicity issues, materials consisting of only naturally occurring components such as polypeptides and polysaccharides can also be used as scaffolds [129]. Often the easiest method for obtaining these biomaterials on scale is directly from biological systems. Scaffolds that are derived from natural sources include collagen, fibronectin, chitosan, and alginate, all of which are constituents of ECM in certain organisms. Products are also available which consist of a natural mixture of substances derived from cells without purification [130]. These include Basement Membrane Extract (BME), also known as Matrigel, which is extracted from murine sarcomas [131]. Such scaffolds often result in favourable responses from cells as they closely resemble the environments that the cells encounter in nature. They can also be degraded by the host organism over time and only contain natural components that can be processed. However, they suffer from disadvantages such as inconsistency between batches, and not being able to determine the precise make-up of the material can lead to unintended cellular responses [132].

The benefits of natural systems can be retained and the shortcomings avoided by using designed synthetic polymers constructed from biological components. The chemical synthesis or biological expression of designed peptide sequences is now well established, and so protein- or peptide-based hydrogels are increasingly being utilised [133–135]. These materials add complete control over scaffold composition alongside the scope to modulate properties and also allow functionalisation. They also usually self-assemble in solution in response to cues such as temperature or pH, which aids handling. Some of the most widely used systems are formed from short synthetic peptides that interact to form entangled fibres of β-sheet [133, 135]. There are potential issues with the degradation of such scaffolds, however, as β-forming peptides can form insoluble amyloids that accumulate in the body and are implicated in numerous diseases [136]. They are also generally only short peptides, with most of the amino-acid residues dedicated to intermolecular assembly, and so offer limited scope for modification.

1.4.2 Hydrogels Derived from α-Helical Peptides

The coiled coil is a structural motif that occurs frequently in natural proteins and is well studied and understood. It consists of α-helical polypeptides coiled around each other in a rope-like structure [137, 138]. Identification of the sequence patterns that lead to coiled-coil assembly has allowed the *de novo* design of synthetic peptides that spontaneously self-assemble into coiled coils in aqueous solution [138–140]. While larger assemblies are becoming better understood and more accessible [141, 142], and can achieve advanced function [143], dimeric coiled coils are the most common and well-developed of all potential coiled-coil assemblies. The archetypal dimeric coiled coil is the 'leucine zipper', with assembly driven by hydrophobic stripes of Leu residues along one side of each α-helix (Fig. 1.11) [144].

The association of coiled coils has been used as the basis for several designed materials [145]. Often dimeric coiled-coil domains serve to cross-link long polypeptide chains, for example in a system that forms a hydrogel reversibly in response to pH and temperature changes [146]. Despite the success of these approaches, the use of relatively large recombinant proteins makes the exact behaviour difficult to characterise, and fine-tuning the materials a challenge.

Hydrogelating Self-Assembling Fibres

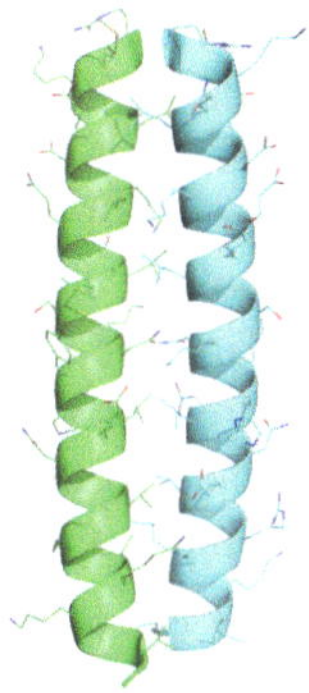

Fig. 1.11 Structure of leucine zipper heterodimeric coiled coil. Image generated from PDB ID 2ZTA using PyMOL [59]

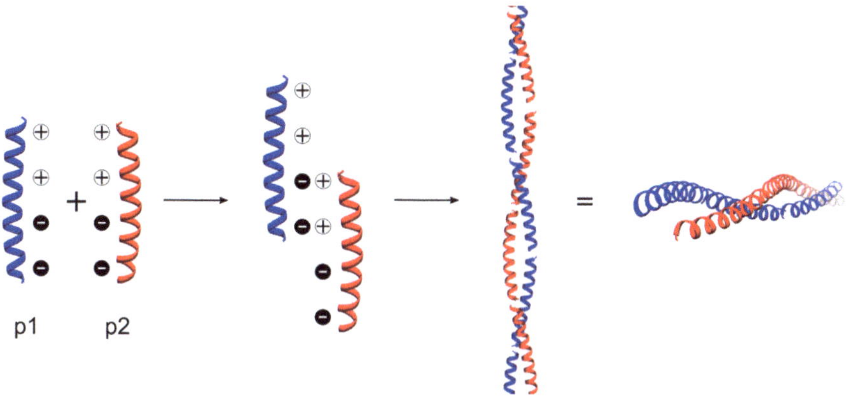

Fig. 1.12 Cartoon representation of self-assembly of fibres from peptides based upon heterodimeric coiled coils

Self-assembling fibres (SAFs) are based on short (28 residue) synthetic peptides that form heterodimeric coiled coils upon mixing which then assemble into μm scale fibres [147]. The key design feature is the staggering of the dimer interface, so that the two peptides assemble in an offset manner leaving 'sticky ends' with unsatisfied interfaces that each bind further peptides and continue to assemble longitudinally (Fig. 1.12). These fibrils assemble further laterally to form larger fibres, the dimensions and properties of which can be altered by changing the amino acid sequences of the peptides [148]. Hydrogelating Self-Assembling Fibres (hSAFs) are formed from peptides based on SAFs that form hydrogels upon mixing, making them an appropriate scaffold for tissue engineering [149].

The constituent peptides of coiled coils can be described by a minimal heptad repeat (of seven amino acids) [137]. This follows from the 3.6 residue per turn of an α-helix, which means that the n and $n+7$ amino acids in the sequence occupy almost equivalent positions. The difference between seven and the optimal two-turn repeat of 7.2 leads to super-coiling of the coiled coil.

The design features that lead to the hSAF peptides assembling as coiled-coil heterodimers are common to all dimeric coiled coils (Fig. 1.13). The structure can be represented as a 'helical wheel', looking down the length of the fibrils along the interface between the two peptides (Fig. 1.13a). Dimeric coiled coils have a heptad repeat of *hpphppp*, where *h* is a hydrophobic residue and *p* polar. The heptad is conventionally labelled *a* through *g*, and so the hydrophobic residues are at the *a* and *d* positions [139]. Upon folding into an α-helix, the hydrophobic side chains are adjacent, forming a hydrophobic stripe down one face of the helix. Burying this stripe is what drives dimerisation in water. For the hSAF design, branched non-polar Leu and isoleucine (Ile) residues are placed at the *a* and *d* positions. These favour parallel assembly through interlocking of the side chains in so-called knobs-into-holes interactions [150] (Fig. 1.13b).

Homodimerisation of the peptides before the two are mixed is prevented by the placement of polar asparagine (Asn) residues at complementary *a* positions in the hydrophobic stripes. These disrupt the hydrophobic assembly, except in the desired staggered heterodimers where the two Asn side chains can form an inter-helix hydrogen bond. The dimerisation interactions are reinforced by complementary charged residues (at neutral pH), in the form of basic lysine (Lys) and acidic glutamic acid (Glu), at the *e* and *g* positions (Fig. 1.13c).

Only four of the seven residues in a coiled-coil dimer heptad repeat are required to form the interface. The difference for hSAFs from SAFs is that the majority of the remaining residues on the outside face away from the dimer interface (*b*, *c*, and *f* positions) are weakly hydrophobic alanine (Ala). This drives lateral association between the coiled-coil fibres in aqueous buffers to form the hydrogel. Some of these outside positions can also be altered if desired, for example with a Trp; at least one Trp at an *f*-position is necessary for the stability of the gels in phosphate-buffered saline (PBS) buffer [149]. The substitution of further *f*-position Ala residues for Trp gives gels of increased stiffness [151].

1.4.3 Functionalisation of Tissue Engineering Scaffolds

In nature, the ECM not only influences cells through its physical properties, but it also contains a myriad of molecular cues that interact with numerous receptors on the cell surface to direct development [119]. These cues can be constituents of the ECM, or reach cells as free species that diffuse through extracellular space. Features of the ECM that influence cellular development include short peptide sequences as well as glycans or glycoconjugates [152, 153]. One of the best understood is the RGD (arginine (Arg)-glycine (Gly)-aspartic acid (Asp)) tripeptide sequence [154]. This short motif is the recognition sequence for integrins, the membrane-bound proteins that anchor cells to their environment, in several polypeptides found in the ECM.

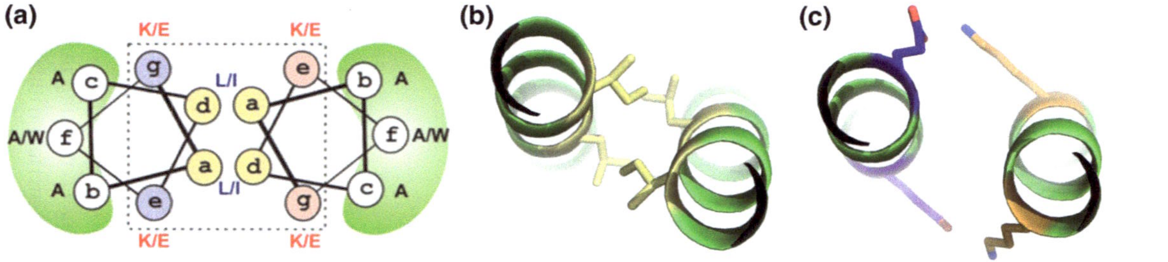

Fig. 1.13 **a** Helical wheel representation of hSAF dimeric coiled-coil interface viewed along the long fibre axis. Single letter codes shown for amino acids at each position. Cartoon representations of interactions between **b** hydrophobic residues at *a* and *d* positions, and **c** polar residues at *e* and *g* positions

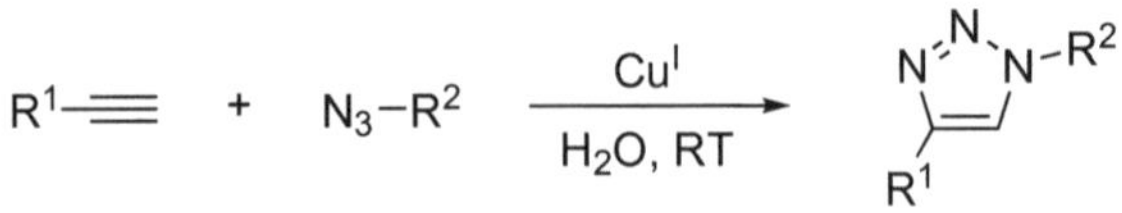

Scheme 1.1 CuAAC reaction for covalently linking two molecules

The performance of tissue engineering scaffolds can be greatly increased by combining the desirable physical properties outlined above with functionalisation with motifs that mimic the biological cues of the ECM appropriate for the tissue [133, 153, 155]. Many techniques exist to allow the functionalisation of materials used as scaffolds with various different moieties, including small molecules, bioactive peptides, and growth factors [156]. The RGD peptide is the most often used to enhance adhesion of cells to the base material [157].

Modification of Scaffolds by 'Click' Chemistry

A common technique for the covalent linkage of molecules is Cu-catalyzed azide-alkyne [3 + 2] cycloaddition (CuAAC) (Scheme 1.1) [158, 159]. It is a reaction that proceeds rapidly and efficiently under mild conditions, thus exemplifying 'click' chemistry [160]. It has become a standard method for modifying materials to introduce biological cues [161, 162], requiring functionalisation of the material with one of the azide or alkyne moieties and the modifiers having the other. Functionalisation with these groups can be achieved for polypeptide-based materials by the incorporation of one of several known non-canonical amino acids, such as azidonorleucine (Anl) [163, 164]. The ubiquity of CuAAC also means that chemical syntheses of many biological motifs with the appropriate functionality have already been established.

1.4.4 Carbohydrates in Tissue Engineering

Even though carbohydrates fulfil many important biological roles (Sect. 1.2.1), they are relatively under-used in the field of tissue engineering. This is despite their potential to modulate the properties of scaffolds as well as stimulate responses from cells. Carbohydrates are key components of the ECM, where they play structural and signalling roles [12, 153, 165]. Several naturally derived polysaccharides, such as alginate, are used as biomaterials for tissue engineering applications with little or no modification [166]. Hybrid systems comprising mixtures of saccharides and other components are also able to form materials with advanced properties, such as high elasticity and toughness [167].

Glycopolymers as Tissue Engineering Scaffolds

Glycosylation of polymeric materials that can be identically functionalised multiple times gives glycopolymers, which can target multivalent carbohydrate-based interactions [168]. Glycopolymers have previously been synthesised for several applications, including drug delivery and inhibiting viral infection [169, 170]. Many are based upon organic polymers, with simple sugar moieties such as galactose appended either to the monomer before polymerisation or to the polymer after formation. This is usually carried out to 'improve biocompatibility' for the unnatural materials, without any specific interactions being targeted, with effects assessed by simple cell culture experiments.

CuAAC reactions are increasingly utilised for the covalent linkage of carbohydrates [171], and have been applied to the synthesis of glycopolymers [172]. They have been used to covalently link carbohydrates onto the appropriately functionalised polymers, meaning that there are known synthetic protocols for simple sugars with azide or alkyne functionality. CuAAC is also sometimes used to assemble the polymer backbone, or to cross-link polymers to form gels.

1.4.4.1 Synthesis of Glycopeptides and Glycoproteins

Many functional glycans in nature exist appended to glycoproteins, and most ECM polypeptides are glycosylated as proteoglycans. Therefore, materials based only upon natural peptide and carbohydrate subunits make attractive targets for biomaterials. Glycopeptides can often mimic many of the key features of natural glycoproteins [173], and can be synthesised more easily than glycoproteins, as the production of proteins as a single desired glycoform remains a considerable challenge [174].

Various methods exist for appending molecules onto proteins to mimic post-translational modifications [175, 176]. These include the chemical or recombinant incorporation of azide-containing amino acids for linkage by CuAAC [177]. This technique has been applied to make synthetic glycopeptides and glycoproteins, both through linkages that occur naturally and artificial ones [178]. Artificial glycosylation of proteins by CuAAC has been used to direct them to sites of inflammation in vivo [179]. Synthetic glycoproteins expand the current scope of the well-developed protein design aspect of synthetic biology to approach the enormous diversity and functionality that nature has successfully evolved with access to only a few molecular building blocks.

In a recent study, fibres formed from the assembly of coiled-coil peptides have been decorated with carbohydrates via amide coupling [180]. This multivalent display of disaccharides from *Leishmania* enhances the binding of antibodies, showing a potential application to medical diagnostics.

Carbohydrates Implicated in Cellular Development

The roles that specific carbohydrates play in biological systems are only beginning to be elucidated. This is due to the complexity of oligosaccharide structures, the resultant difficulties with glycomics (Sect. 1.1.1), and the challenges of making oligosaccharides synthetically (Sect. 1.5). Meanwhile, difficulty in accessing single glycoforms of glycoproteins means that often it is not possible to determine the biological or physical effects caused by specific glycan structures.

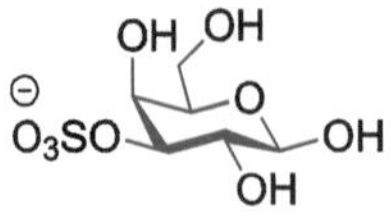

15

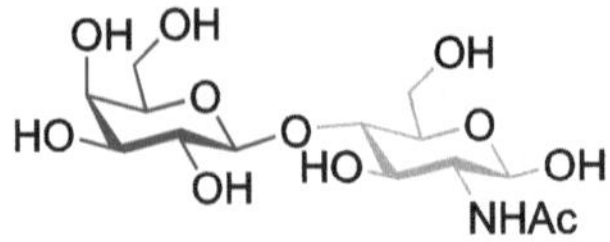

16

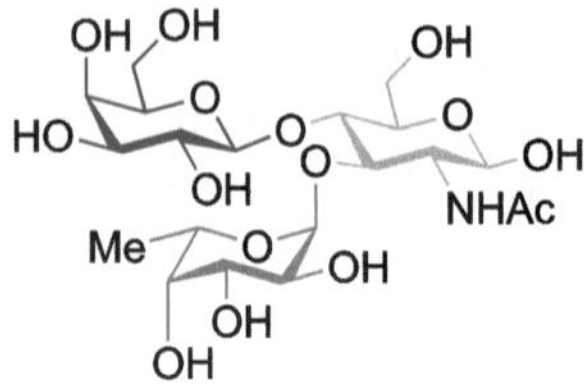

17

Fig. 1.14 Structures of the biologically relevant carbohydrates SGal (**15**), N-acetyllactosamine (LacNAc) (**16**), and Lewis^X (Le^X) (**17**)

A carbohydrate with an observable cellular response relevant to tissue engineering is Gal (**5**). It has long been known that hepatocytes interact with Gal *via* the asialoglycoprotein, which was discovered by covalent attachment of the monosaccharide to surfaces upon which cells were cultured [181]. Gal and its derivatives are often used to decorate scaffolds for liver tissue engineering to enhance cellular function [182].

Gal is also proposed to play a role in the development of oligodendrocytes, the neural cells responsible for forming myelin in the central nervous system, [183] and so is of interest for neural tissue engineering. The cell membranes of oligodendrocytes, and hence the myelin sheaths that they form, are rich in glycolipids, namely galactosylceramide and its 3-sulfated derivative sulfatide [184], therefore presenting an array of Gal and D-galactose-3-sulfate (SGal) (**15**) groups Fig. 1.14. It has been suggested that carbohydrate–carbohydrate interactions between these moieties are important for oligodendrocyte function, as myelin sheaths wrap around nerve axons many times and so have multiple membrane–membrane interfaces [184]. Indeed, oligodendrocytes have a measurable cellular response (de-differentiation) to vesicles or nanoparticles decorated with a combination of Gal and SGal [185]. This agrees with the fact that sulfatide negatively regulates oligodendrocyte differentiation in sulfatide-null mice [186].

Other potential targets for carbohydrates in tissue engineering are galectins, which are cell-surface receptors that bind β-galactosides. Galectins are implicated in several cellular processes [187–189], including: cellular adhesion and proliferation [190]; growth and apoptosis [191]; and endothelial cell vascularisation and morphogenesis [192]. They are also important for neural stem cells, including in neurogenesis [193] and promoting proliferation [194]. Relevant β-galactosides include LacNAc (**16**), which has been shown to bind to galectins [195], and the related trisaccharide Le^x (**17**) Fig. 1.14, which is a biomarker on neural stem cells [196], as well as other cell types [197].

Finally, glycosaminoglycans have recently been shown to influence cellular development in their role as ECM components. Glycosaminoglycans are formed from repeating disaccharide units, consisting of one *N*-acetylamino sugar and one uronic acid (or galactose). They are implicated in angiogenesis [198], and so could be useful in developing vascular tissues, and also neurogenesis [199–201].

1.5 Carbohydrate Synthesis

Synthesis of specific carbohydrates is often the limiting factor in studies involving carbohydrates. The complexity of oligosaccharides, and chemical similarity within and between monosaccharide units, makes their synthesis challenging. Methods for forming specific glycosidic linkages based solely on chemical synthesis are now being improved and replaced by methods utilising enzymes, although some traditional synthetic chemistry is often still required to give the functionality required to make the product useful, such as a chemical handle or fluorescent probe [18, 202].

1.5.1 Chemical Synthesis of Carbohydrates

There have been many impressive achievements by purely chemical synthesis of carbohydrates, including drugs and synthetic vaccines [17, 203]. However, carbohydrate structure makes it a laborious process. To make a single glycosidic linkage requires multi-step syntheses of an appropriate glycosyl donor and the corresponding glycosyl acceptor (Fig. 1.15). This is to ensure reaction at the desired hydroxyl on each glycoside, and precise manipulation of protecting groups is required to ensure the correct positions are activated on the donor, and unprotected on the acceptor. The correct combination of protecting and activating groups is also needed to give the α- or β-linkage as required [204–206].

To form subsequent glycosidic linkages to give higher oligosaccharides requires a protecting group in the starting components that can be orthogonally deprotected or activated. Many methodologies have been developed to facilitate chemical carbohydrate synthesis, including one-pot [207], and automated solid-phase [208] strategies, but these are as yet not robust and general.

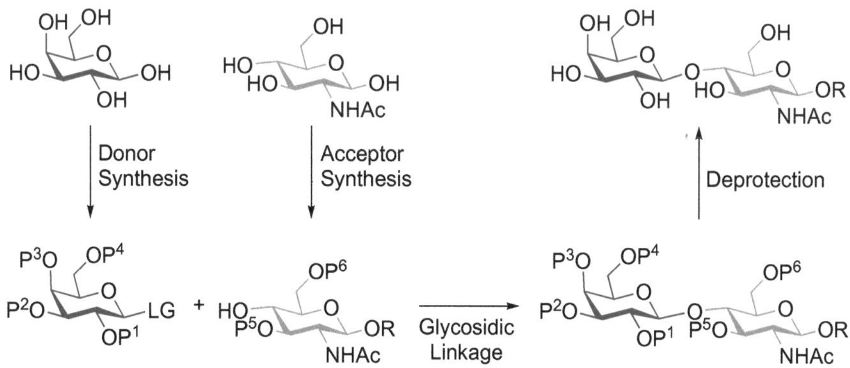

Fig. 1.15 General scheme for chemical synthesis of functionalised disaccharide LacNAc. P = protecting group; LG = leaving group; R = functional group

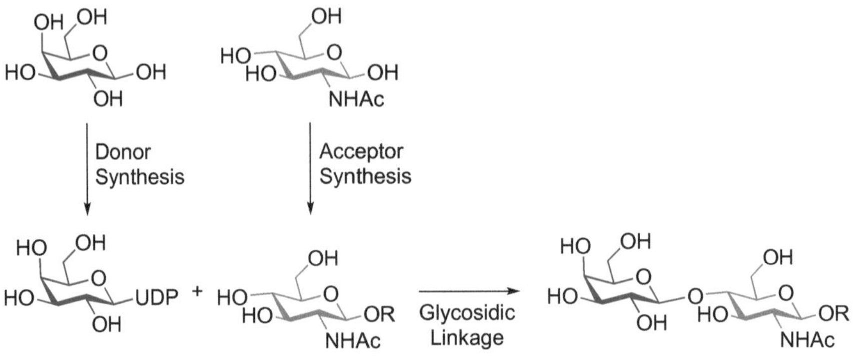

Fig. 1.16 General schematic for enzymatic synthesis of disaccharide LacNAc. R = functional group

1.5.2 *Enzymatic Synthesis of Carbohydrates*

Synthesis of oligosaccharides utilising natural or engineered enzymes eliminates many of the issues associated with chemical synthesis [209]. Enzymes bind the glycosyl donor and acceptor in such a way as to lead only to formation of the glycosidic linkage at the desired position and in the desired anomer (Fig. 1.16). This means that protecting groups are not necessary, greatly reducing the number of required synthetic steps. Substitution at the anomeric position of the glycosyl acceptor is generally tolerated, and so monosaccharides with desired functionality can be synthesised and built up into oligosaccharides enzymatically. The reactions also proceed in aqueous media at or near room temperature and neutral pH.

The disadvantages of enzymatic couplings include specificity; there are only a few glycosidic linkages for which appropriate enzymes have been identified and isolated in a form that allows application to synthesis. Even for those that have been identified, obtaining sufficient protein for reactions above small-scale is often difficult, and the glycosyl donors can be costly to make. Enzymes are sensitive to reaction conditions, generally requiring substrates that are water soluble and stable under biological conditions.

Glycosyltransferases in Oligosaccharide Synthesis

Glycosyltransferases are the enzymes that perform glycosylation in nature, with acceptors ranging from carbohydrates to proteins to lipids. These enzymes are therefore ideally suited for enzymatic oligosaccharide synthesis when appropriate examples can be identified and isolated, if conditions can be found under which they are active and stable [210]. This has been achieved for galactosyl-[211], fucosyl-[212], and sialyltransferases [213], amongst others.

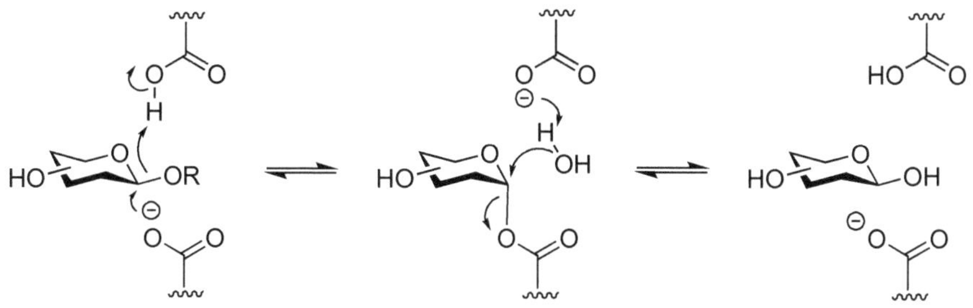

Scheme 1.2 Mechanism of glycosidic bond hydrolysis by a typical retaining glycosidase. Carboxylic acid groups represent side chains of catalytic Asp/Glu residues in enzyme active site. This process is an equilibrium, so can be driven towards bond formation under appropriate conditions

The donors for glycosyltransferases, nucleoside-phosphate glycosides, can be made enzymatically and many are available commercially. The reactions can be driven to completion by addition of phosphatases, which hydrolyse the nucleoside-phosphate by-products [214]. Specific multivalent cations usually coordinate the phosphate groups in the active sites, and so are required additives. Substrate specificity can be broadened by rational mutations [215] or directed evolution [216].

Glycosynthases

For some desired glycosidic linkages, appropriate glycosyltransferases have not been identified, or isolated in a useable form, and so alternative enzymes must be found or developed. As enzymes catalyse reactions in both the forward and reverse direction, one option is to use glycosidases. In nature these enzymes serve to cleave glycosidic linkages (Scheme 1.2), but as this process is an equilibrium under the appropriate conditions they can be used to build oligosaccharides [217], including as glycans appended to proteins [218].

In some cases, engineering of the glycosidases is required to give efficient bond-forming reactions (Scheme 1.3). The resultant enzymes, termed glycosynthases, have been pioneered by the Withers group [219, 220]. The archetypal glycosynthase is a mutant of a galacto-/glucosidase, and so can add Gal or Glc residues onto gluco-acceptors [221]. As with glycosyltransferases, it is possible to greatly increase the substrate scope through directed evolution of the the first rationally designed mutants [222].

Glycosynthases are developed from glycosidases by removal of the key catalytic residue, which cleaves the glycosidic bond, from the active site. Use of an appropriate donor that mimics the covalent intermediate of the cleavage, often an α-fluoro glycoside [223], allows for formation rather than breaking of this linkage.

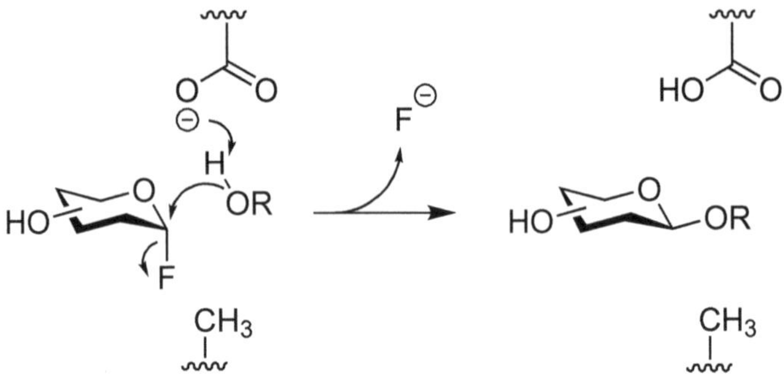

Scheme 1.3 Mechanism of glycosidic bond formation by glycosynthase engineered from glycosidase. Mutation of a key residue to Ala (side chain represented as a methyl group) prevents bond hydrolysis. α-Glycosyl fluorides mimic the glycosidase transition states and can act as donors in glycosylation reactions

1.6 Scope of the Thesis Part II: Tissue Engineering Scaffolds Functionalised with Carbohydrates

The second part of this thesis describes the development of enhanced scaffolds for tissue engineering applications based upon a modifiable derivative of the hSAF system functonalised with carbohydrates. hSAFs are an ideal system to use as a scaffold for tissue engineering, forming hydrogels with desirable physical properties under biocompatible conditions. They are made up only of peptides, which are easy to handle and are biological units and so should be degraded to harmless by-products in the body over time. Being synthetic means that the constitution of the scaffold is known and precisely controlled, and producing derivatives to enhance function is straightforward. The coiled coil system itself is robust and amenable to alteration.

The utility of the hSAFs is enhanced by making them modifiable, i.e., designing a version to which functional biological motifs can be appended. This allows the tuning of the scaffold for particular tissue types, e.g., by the appending of cues to direct stem cells to differentiate into particular lineages. Before the research described here, a derivative of the hSAFs system had been developed within the Woolfson group with an azide-containing amino acid, which can be functionalised with alkyne-containing motifs by CuAAC. This system was used as a basis for this work, as the ubiquity of copper 'click' chemistry means a wide range of appropriate motifs are accessible. Also, azide-containing amino acids are compatible with recombinant expression of the peptides, which might be required for eventual large-scale production of the functionalised system.

Carbohydrates are currently under-utilised in tissue engineering, due to a range of synthetic challenges, but have the potential to be useful given the wide range of roles that they play in biology. The modifiable hSAFs provide an appropriate scaffold for modification with carbohydrates as the ligation of carbohydrates to polypeptides is well-established, including by CuAAC. The fibres can be functionalised multiply with carbohydrate moieties to target multivalent carbohydrate-based interactions. I used chemical synthesis to make a library of alkyne-functionalised monosaccharides, and with these to verify the modification of the azide-containing hSAF derivative system with carbohydrates, ensuring that the hydrogel structure and properties are not altered unfavourably.

Many of the carbohydrates implicated in cellular development have effects on neural cell types. These were primarily targeted, as the hSAFs are of the approximate stiffness of neural tissue. As well as monosaccharides such as Gal and SGal, a library of oligosaccharides with known biological responses was synthesised using chemoenzymatic techniques. This library included β-galactosides, including LacNAc and Le^X, which may be of use for neural tissue engineering.

References

1. Walt D, Aoki-Kinoshita KF, Bendiak B, Bertozzi CR, Boons G-J, Darvill A, Hart G, Kiessling LL, Lowe J, Moon RJ, Paulson JC, Sasisekharan R, Varki AP, Wong C-H (2012) Transforming glycoscience: a roadmap for the future. National Academies Press, Washington, D.C
2. Wang L-X, Davis BG (2013) Chem Sci 4:3381–3394
3. Varki A (1993) Glycobiology 3:97–130
4. Dwek RA (1996) Chem Rev 96:683–720
5. Varki A, Cummings RD, Esko JD, Freeze HH, Stanley P, Bertozzi CR, Hart GW, Etzler ME (eds) (2009) Essentials of glycobiology, 2nd edn. Cold Spring Harbor Laboratory Press, Cold Spring Harbor (NY)
6. Dalziel M, Crispin M, Scanlan CN, Zitzmann N, Dwek RA (2014) Science 343:1235681
7. Walsh CT, Garneau-Tsodikova S, Gatto GJ (2005) Angew Chem Int Ed 44:7342–7372
8. Davis BG (2009) Pure Appl Chem 81:285–298
9. Solís D, Bovin NV, Davis AP, Jiménez-Barbero J, Romero A, Roy R, Smetana K, Gabius H-J (1850) Biochim Biophys Acta 2015:186–235
10. Shental-Bechor D, Levy Y (2009) Curr Opin Struct Biol 19:524–533
11. Gandhi NS, Mancera RL (2008) Chem Biol Drug Des 72:455–482
12. Toole BP (2004) Nat Rev Cancer 4:528–539
13. Cremer D, Pople JA (1975) J Am Chem Soc 97:1354–1358
14. Bohne-Lang A, Lang E, Förster T, von der Lieth C-W (2001) Carbohydr Res 336:1–11
15. Ambrosi M, Cameron NR, Davis BG (2005) Org Biomol Chem 3:1593–1608
16. Hebert DN, Lamriben L, Powers ET, Kelly JW (2014) Nat Chem Biol 10:902–910
17. Ernst B, Magnani JL (2009) Nat Rev Drug Discov 8:661–677
18. Kiessling LL, Splain RA (2010) Annu Rev Biochem 79:619–653
19. Berti F, Adamo R (2013) ACS Chem Biol 8:1653–1663
20. Paszek MJ et al (2014) Nature 511:319–325
21. Hanson SR, Culyba EK, Hsu T-L, Wong C-H, Kelly JW, Powers ET (2009) Proc Natl Acad Sci USA 106:3131–3136
22. Wormald MR, Dwek RA (1999) Structure 7:155–160
23. Cummings RD, Pierce JM (2014) Chem Biol 21:1–15
24. Grant OC, Woods RJ (2014) Curr Opin Struct Biol 28C:47–55
25. Pérez S, Tvaroška I (2014) In: Horton D (ed) Advances in carbohydrate chemistry and biochemistry, vol 71, 1st edn. Academic Press, London, pp 9–136
26. Johnson Q, Lindsay R, Petridis L, Shen T (2015) Molecules 20:7700–7718
27. Mareška V, Tvaroška I, Králová B, Spiwok V (2015) Carbohydr Res 408:1–7

28. Fernández-Alonso MC, Díaz D, Berbis MÁ, Marcelo F, Cañada J, Jiménez-Barbero J (2012) Curr Protein Pept Sci 13:816–830
29. Martín-Santamaría S, Gabius H-J, Jiménez-Barbero J (2012) Pure Appl Chem 84:49–64
30. Arnaud J, Audfray A, Imberty A (2013) Chem Soc Rev 42:4798–4813
31. Berman HM, Westbrook J, Feng Z, Gilliland G, Bhat TN, Weissig H, Shindyalov IN, Bourne PE (2000) Nucl Acids Res 28:235–242
32. Berman HM, Coimbatore Narayanan B, Di Costanzo L, Dutta S, Ghosh S, Hudson BP, Lawson CL, Peisach E, Prlić A, Rose PW, Shao C, Yang H, Young J, Zardecki C (2013) FEBS Lett 587:1036–1045
33. Jo S, Song KC, Desaire H, MacKerell AD, Im W (2011) J Comput Chem 32:3135–3141
34. Dunitz JD, Gavezzotti A (2005) Angew Chem Int Ed 44:1766–1787
35. Lütteke T, Frank M, von der Lieth C-W (2005) Nucl Acids Res 33:D242–D246
36. Lütteke T, Frank M, von der Lieth C-W (2004) Carbohydr Res 339:1015–1020
37. Campbell MP et al (2014) BMC Bioinf 15:S9
38. Cantarel BL, Coutinho PM, Rancurel C, Bernard T, Lombard V, Henrissat B (2009) Nucl Acids Res 37:D233–D328
39. Taroni C, Jones S, Thornton JM (2000) Protein Eng 13:89–98
40. Malik A, Ahmad S (2007) BMC Struct Bol 7:1
41. Santos JCA, Nassif H, Page D, Muggleton SH, Sternberg MJE (2012) BMC Bioinf 13:162
42. Tsai K-C, Jian J-W, Yang E-W, Hsu P-C, Peng H-P, Chen C-T, Chen J-B, Chang J-Y, Hsu W-L, Yang A-S (2012) PLoS One 7:e40846
43. Nishio M (2011) Phys Chem Chem Phys 13:13873–13900
44. Abbott DW, van Bueren AL (2014) Curr Opin Struct Biol 28C:32–40
45. Sujatha MS, Balaji PV (2004) Proteins 55:44–65
46. Bhagavat R, Chandra N (2014) Glycobiology 24:5–16
47. Boraston AB, Bolam DN, Gilbert HJ, Davies GJ (2004) Biochem J 382:769–781
48. Petrescu A-J, Milac A-L, Petrescu SM, Dwek RA, Wormald MR (2004) Glycobiology 14:103–14
49. Widmalm G (2013) Carbohydr Res 378:123–32
50. Taylor ME, Drickamer K (2014) Curr Opin Struct Biol 28C:14–22
51. Mammen M, Choi S-K, Whitesides GM (1998) Angew Chem Int Ed 37:2754–2794
52. Reynolds M, Pérez SCR (2011) Chim 14:74–95
53. Lundquist JJ, Toone EJ (2002) Chem Rev 102:555–578
54. Quiocho F (1986) Annu Rev Biochem 55:287–315
55. Vyas NK (1991) Curr Opin Struct Biol 1:732–740
56. Weis WI, Drickamer K (1996) Annu Rev Biochem 65:441–473
57. Chakrabarti S, Hinczewski M, Thirumalai D (2014) Proc Natl Acad Sci USA 111:9048–9053
58. Binder FPC, Lemme K, Preston RC, Ernst B (2012) Angew Chem Int Ed 51:7327–7331
59. The PyMOL Molecular Graphics System, Version 1.7.0.3. (2014)
60. Persch E, Dumele O, Diederich F (2015) Angew Chem Int Ed 54:3290–3327
61. Asensio JL, Ardá A, Cañada FJ, Jiménez-Barbero J (2013) Acc Chem Res 46:946–954
62. Muraki M, Harata K, Sugita N, Sato K-I (2000) Biochemistry 101:292–299
63. Meyer JE, Schulz GE (1997) Protein Sci 6:1084–1091
64. Bernardi A, Arosio D, Potenza D, Sánchez-Medina I, Mari S, Cañada FJ, Jiménez-Barbero J (2004) Chem - Eur J 10:4395–4406
65. Culyba EK, Price JL, Hanson SR, Dhar A, Wong C-H, Gruebele M, Powers ET, Kelly JW (2011) Science 331:571–5
66. Vacas T, Corzana F, Jiménez-Osés G, González C, Gómez AM, Bastida A, Revuelta J, Asensio JL (2010) J Am Chem Soc 132:12074–12090
67. Mullins EA, Shi R, Parsons ZD, Yuen PK, David SS, Igarashi Y, Eichman BF (2015) Nature 527:254–258
68. Vyas NK, Vyas MN, Quiocho FA (1988) Science 242:1290–1295
69. Ferrand Y, Crump MP, Davis AP (2007) Science 318:619–622
70. Hanson QM, Barb AW (2015) Biochemistry 54:2931–2942

71. Hang I, Lin C-W, Grant OC, Fleurkens S, Villiger TK, Soos M, Morbidelli M, Woods RJ, Gauss R, Aebi M (2015) Glycobiology 25:1335–1349
72. Murray AN, Chen W, Antonopoulos A, Hanson SR, Wiseman RL, Dell A, Haslam SM, Powers DL, Powers ET, Kelly JW (2015) Chem Biol 22:1052–1062
73. Fauchere JL, Pliska VE (1983) Eur J Med Chem 18:369–375
74. Lemieux RU (1996) Acc Chem Res 4842:373–380
75. Stanca-Kaposta EC, Gamblin DP, Screen J, Liu B, Snoek LC, Davis BG, Simons JP (2007) Phys Chem Chem Phys 9:4444–4451
76. Hunter CA, Lawson KR, Perkins J, Urch CJ (2001) J Chem Soc Perkin Trans 2:651–669
77. Salonen LM, Ellermann M, Diederich F (2011) Angew Chem Int Ed 50:4808–4842
78. Nishio M, Umezawa Y, Fantini J, Weiss MS, Chakrabarti P (2014) Phys Chem Chem Phys 16:12648–12683
79. Brandl M, Weiss MS, Jabs A, Sühnel J, Hilgenfeld R (2001) J Mol Biol 307:357–377
80. Plevin MJ, Bryce DL, Boisbouvier J (2010) Nat Chem 2:466–471
81. Suezawa H, Hashimoto T, Tsuchinaga K, Yoshida T, Yuzuri T, Sakakibara K, Hirota M, Nishio M (2000) J Chem Soc Perkin Trans 2:1243–1249
82. Zondlo NJ (2013) Acc Chem Res 46:1039–1049
83. Ams MR, Fields M, Grabnic T, Janesko BG, Zeller M, Sheridan R, Shay A (2015) J Org Chem 80:7764–7769
84. Huang K-Y, Horng J-C (2015) Biochemistry 54:6186–6194
85. Nikolova V, Ilieva S, Galabov B, Schaefer HF III (2014) J Org Chem 79:6823–6831
86. Tsuzuki S, Fujii A (2008) Phys Chem Chem Phys 10:2584–2594
87. Bloom JWG, Raju RK, Wheeler SE (2012) J Chem Theory Comput 8:3167–3174
88. Mishra BK, Karthikeyan S, Ramanathan V (2012) J Chem Theory Comput 8:1935–1942
89. Mishra BK, Deshmukh MM, Venkatnarayan R (2014) J Org Chem 79:8599–8606
90. Wheeler SE (2012) Acc Chem Res 46:1029–1038
91. Parrish RM, Sherrill CD (2014) J Am Chem Soc 136:17386–17389
92. Cockroft SL, Perkins J, Zonta C, Adams H, Spey SE, Low CMR, Vinter JG, Lawson KR, Urch CJ, Hunter CA (2007) Org Biomol Chem 5:1062–1080
93. Hunter CA, Low CMR, Rotger C, Vinter JG, Zonta C (2002) Proc Natl Acad Sci USA 99:4873–4876
94. Dougherty DA (2013) Acc Chem Res 46:885–893
95. Mahadevi AS, Sastry GN (2013) Chem Rev 113:2100–2138
96. Gomes JANF, Mallion RB (2001) Chem Rev 101:1349–1383
97. Fernández-Alonso MdC, Cañada FJ, Jiménez-Barbero J, Cuevas G (2005) J Am Chem Soc 127: 7379–7386
98. Jiménez-Moreno E, Jiménez-Osés G, Gómez AM, Santana AG, Corzana F, Bastida A, Jiménez-Barbero J, Asensio JL (2015) Chem Sci 6:6076–6085
99. Laughrey ZR, Kiehna SE, Riemen AJ, Waters ML (2008) J Am Chem Soc 130:14625–14633
100. Barwell NP, Davis AP (2011) J Org Chem 76:6548–6557
101. Chen W, Enck S, Price JL, Powers DL, Powers ET, Wong C-H, Dyson HJ, Kelly JW (2013) J Am Chem Soc 135:9877–9884
102. Lucas R, Peñalver P, Gómez-Pinto I, Vengut-Climent E, Mtashobya L, Cousin J, Maldonado OS, Perez V, Reynes V, Aviñó A, Eritja R, González C, Linclau B, Morales JC (2014) J Org Chem 79:2419–2429
103. Wohlert J, Schnupf U, Brady JW (2010) J Chem Phys 133:155103
104. Tsuzuki S, Uchimaru T, Mikami M (2011) J Phys Chem A 113:11256–11262
105. Kumari M, Sunoj RB, Balaji PV (2012) Org Biomol Chem 10:4186–4200
106. Wimmerová M, Kozmon S, Nečasová I, Mishra SK, Komárek J, Koča J (2012) PLoS One 7:e46032
107. Kozmon S, Matuška R, Spiwok V, Koča J (2011) Phys Chem Chem Phys 13:14215–14222
108. Langer R, Vacanti J (1993) Science 260:920–926
109. van Blitterswijk C, Thomsen P, Lindahl A, Hubbell J, Williams DF, Cancedda R, de Bruijn JD, Sohier J (eds.) (2008) Tissue engineering. Academic Press

110. Mao AS, Mooney DJ (2015) Proc Natl Acad Sci USA 112:14452–14459
111. George JH (2009) Engineering of fibrous scaffolds for use in regenerative medicine. Doctoral Thesis, Imperial College, London
112. Atala A, Bauer SB, Soker S, Yoo JJ, Retik AB (2006) Lancet 367:1241–1246
113. Macchiarini P, Jungebluth P, Go T, Asnaghi MA, Rees LE, Cogan TA, Dodson A, Martorell J, Bellini S, Parnigotto PP, Dickinson SC, Hollander AP, Mantero S, Conconi MT, Birchall MA (2008) Lancet 372:2023–2030
114. Bianco P, Robey PG (2001) Nature 414:118–121
115. Discher DE, Mooney DJ, Zandstra PW (2009) Science 324:1673–1677
116. Dingal PCDP, Discher DE (2014) Nat Mater 13:532–537
117. Discher DE, Janmey P, Wang Y-L (2005) Science 310:1139–1143
118. Stevens MM, George JH (2005) Science 310:1135–1138
119. Juliano RL, Haskill S (1993) J Cell Biol 120:577–585
120. Lutolf MP, Hubbell JA (2005) Nat Biotechnol 23:47–55
121. Tibbitt MW, Rodell CB, Burdick JA, Anseth KS (2015) Proc Natl Acad Sci USA 112:14444–14451
122. Nair LS, Laurencin CT (2007) Prog Polym Sci 32:762–798
123. Rubashkin MG, Ou G, Weaver VM (2014) Biochemistry 53:2078–2090
124. Engler AJ, Sen S, Sweeney HL, Discher DE (2006) Cell 126:677–689
125. Wen JH, Vincent LG, Fuhrmann A, Choi YS, Hribar KC, Taylor-weiner H, Chen S, Engler AJ (2014) Nat Mater 13:979–987
126. Slaughter BV, Khurshid SS, Fisher OZ, Khademhosseini A, Peppas NA (2009) Adv Mater 21:3307–3329
127. Seliktar D (2012) Science 336:1124–1128
128. Shi J, Xing MMQ, Zhong W (2012) Membranes (Basel, Switz.) 2012(2): 70–90
129. Van Vlierberghe S, Dubruel P, Schacht E (2011) Biomacromolecules 12:1387–1408
130. Fitzpatrick LE, Mcdevitt TC (2014) Biomater Sci 3:12–24
131. Kleinman HK (2001) Curr Protoc Cell Biol 2001: 00:10.3:10.3.1–10.3.9
132. Li Q, Uygun BE, Geerts S, Ozer S, Scalf M, Gilpin SE, Ott HC, Yarmush ML, Smith LM, Welham NV, Frey BL (2015) Biomaterials 75:37–46
133. Kopeček J, Yang J (2012) Angew Chem Int Ed 51:7396–7417
134. Hosseinkhani H, Hong P-D, Yu D-S (2013) Chem Rev 113:4837–4861
135. Rubert Pérez CM, Stephanopoulos N, Sur S, Lee SS, Newcomb C, Stupp S (2014) I. Ann Biomed Eng 43:501–514
136. Eisenberg D, Jucker M (2012) Cell 148:1188–1203
137. Lupas AN, Gruber M (2005) Adv Protein Chem 70:37–78
138. Parry DAD, Fraser RDB, Squire JM (2008) J Struct Biol 163:258–269
139. Woolfson DN (2005) Adv Protein Chem 70:79–112
140. Fletcher JM, Boyle AL, Bruning M, Bartlett GJ, Vincent TL, Zaccai NR, Armstrong CT, Bromley EHC, Booth PJ, Brady RL, Thomson AR, Woolfson DN (2012) ACS Synth Biol 1:240–250
141. Woolfson DN, Bartlett GJ, Bruning M, Thomson AR (2012) Curr Opin Struct Biol 22:432–441
142. Thomson AR, Wood CW, Burton AJ, Bartlett GJ, Sessions RB, Brady RL, Woolfson DN (2014) Science 346:485–488
143. Burton AJ, Thomas F, Agnew C, Hudson KL, Halford SE, Brady RL, Woolfson DN (2013) J Am Chem Soc 135:12524–12527
144. O'Shea EK, Klemm JD, Kim PS, Alber T (1991) Science 254:539–544
145. Quinlan RA, Bromley EH, Pohl E (2015) Curr Opin Cell Biol 32:131–137
146. Petka WA, Harden JL, McGrath KP, Wirtz D, Tirrell DA (1998) Science 281:389–392
147. Pandya MJ, Spooner GM, Sunde M, Thorpe JR, Rodger A, Woolfson DN (2000) Biochemistry 39:8728–8734
148. Ryadnov MG, Woolfson DN (2003) Nat Mater 2:329–332

149. Banwell EF, Abelardo ES, Adams DJ, Birchall MA, Corrigan A, Donald AM, Kirkland M, Serpell LC, Butler MF, Woolfson DN (2009) Nat Mater 8:596–600
150. Hadley EB, Testa OD, Woolfson DN, Gellman SH (2008) Proc Natl Acad Sci USA 105:530–535
151. Wasmuth A, Mehrban N, Hudson KL, Thomson AR, Mullen LM, Birchall MA, Woolfson DN Manuscript in preparation
152. Pradhan S, Farach-Carson MC (2010) Regen Med 5:961–970
153. Custódio CA, Reis RL, Mano JF (2014) Adv Healthc Mater 3:797–810
154. Pierschbacher MD, Ruoslahti E (1984) Nature 309:30–33
155. Woolfson DN, Mahmoud ZN (2010) Chem Soc Rev 39:3464–3479
156. Debets MF, van Hest JCM, Rutjes FPJT (2013) Org Biomol Chem 11:6439–6455
157. Hersel U, Dahmen C, Kessler H (2003) Biomaterials 24:4385–4415
158. Rostovtsev VV, Green LG, Fokin VV, Sharpless KB (2002) Angew Chem Int Ed 41:2596–2599
159. Hein JE, Fokin VV (2010) Chem Soc Rev 39:1302–1315
160. Kolb HC, Finn MG, Sharpless KB (2001) Angew Chem Int Ed 40:2004–2021
161. Lutz J-F, Zarafshani Z (2008) Adv Drug Deliv Rev 60:958–970
162. Golas PL, Matyjaszewski K (2010) Chem Soc Rev 39:1338–1354
163. Beatty K, Tirrell DA (2009) Protein Eng 2009: 127–153
164. Voloshchuk N, Montclare JK (2010) Mol Biosyst 6:65–80
165. Bishop JR, Schuksz M, Esko JD (2007) Nature 446:1030–1037
166. Shelke NB, James R, Laurencin CT, Kumbar SG (2014) Polym Adv Technol 25:448–460
167. Sun J-Y, Zhao X, Illeperuma WRK, Chaudhuri O, Oh KH, Mooney DJ, Vlassak JJ, Suo Z (2012) Nature 489:133–136
168. Becer CR (2012) Macromol Rapid Commun 33:742–752
169. Spain SG, Cameron NR (2011) Polym Chem 2:60–68
170. Miura Y, Hoshino Y, Seto H (2015) Chem Rev, In Press. https://doi.org/10.1021/acs.chemrev.5b00247
171. Dondoni A (2007) Chem - Asian J 2:700–708
172. Slavin S, Burns J, Haddleton DM, Becer CR (2011) Eur Polym J 47:435–446
173. Bonduelle C, Lecommandoux S (2013) Biomacromolecules 14:2973–2983
174. Wang P, Dong S, Shieh J-H, Peguero E, Hendrickson R, Moore MAS, Danishefsky SJ (2013) Science 342:1357–1360
175. Carrico IS (2008) Chem Soc Rev 37:1423–1431
176. Hackenberger CPR, Schwarzer D (2008) Angew Chem Int Ed 47:10030–10074
177. Kiick KL, Saxon E, Tirrell DA, Bertozzi CR (2002) Proc Natl Acad Sci USA 99:19–24
178. Gamblin DP, Scanlan EM, Davis BG (2009) Chem Rev 109:131–163
179. van Kasteren SI, Kramer HB, Jensen HH, Campbell SJ, Kirkpatrick J, Oldham NJ, Anthony DC, Davis BG (2007) Nature 446:1105–1109
180. Zacco E, Anish C, Martin CE, Berlepsch HV, Brandenburg E, Seeberger PH, Koksch B (2015) Biomacromolecules 16: 2188–2197
181. Oka JA, Weigel PH (1986) J Cell Biol 103:1055–1060
182. Cho CS, Seo SJ, Park IK, Kim SH, Kim TH, Hoshiba T, Harada I, Akaike T (2006) Biomaterials 27:576–585
183. Baumann N, Pham-Dinh D (2001) Physiol Rev 81:871–927
184. Boggs JM, Gao W, Hirahara Y (2008) Biochim Biophys Acta, Gen Subj 1780: 445–455
185. Boggs JM, Gao W, Zhao J, Park H-J, Liu Y, Basu A (2010) FEBS Lett 584:1771–1778
186. Hirahara Y, Bansal R, Honke K, Ikenaka K, Wada Y (2004) Glia 45:269–277
187. Camby I, Le Mercier M, Lefranc F, Kiss R (2006) Glycobiology 16:137R–157R
188. Dumic J, Dabelic S, Flögel M (2006) Biochim Biophys Acta 1760:616–635
189. Boscher C, Dennis JW, Nabi IR (2011) Curr Opin Cell Biol 23:383–392
190. Perillo NL, Marcus ME, Baum LG (1998) J Mol Med 76:402–412
191. Yang R, Liu F (2003) Cell Mol Life Sci C 60:267–276

192. Nangia-Makker P, Honjo Y, Sarvis R, Akahani S, Hogan V, Pienta KJ, Raz A (2000) Am J Pathol 156:899–909
193. Sakaguchi M, Okano H (2012) Dev Neurobiol 2012: 1059–1067
194. Sakaguchi M et al (2006) Proc Natl Acad Sci USA 103:7112–7117
195. Ahmad N, Gabius H-J, Sabesan S, Oscarson S, Brewer CF (2004) Glycobiology 14:817–825
196. Hennen E, Faissner A (2012) Int J Biochem Cell Biol 44:830–833
197. Muramatsu T, Muramatsu H (2004) Glycoconj J 21:41–45
198. Chwalek K, Tsurkan MV, Freudenberg U, Leibniz CW (2014) Sci Rep 4:4414
199. Swarup VP, Hsiao TW, Zhang J, Prestwich GD, Kuberan B, Hlady V (2013) J Am Chem Soc 135:13488–13494
200. Huang ML, Smith RAA, Trieger GW, Godula K (2014) J Am Chem Soc 136:10565–10568
201. Liu P, Chen L, Toh JKC, Ang YL, Jee J-E, Lim J, Lee SS, Lee S-G (2015) Chem Sci 6:450–456
202. Wang Z, Chinoy ZS, Ambre SG, Peng W, McBride R, de Vries RP, Glushka J, Paulson JC, Boons G-J (2013) Science 341:379–383
203. Galan MC, Benito-Alifonso D, Watt GM (2011) Org Biomol Chem 9:3598–3610
204. Boltje TJ, Buskas T, Boons G-J (2009) Nat Chem 1:611–622
205. Zhu X, Schmidt RR (2009) Angew Chem Int Ed 48:1900–1934
206. Ranade SC, Demchenko AV (2013) J Carbohydr Chem 32:1–43
207. Hsu Y, Lu X-A, Zulueta MML, Tsai C-M, Lin K-I, Hung S-C, Wong C-H (2012) J Am Chem Soc 134:4549–4552
208. Plante OJ, Palmacci ER, Seeberger PH (2001) Science 291:1523–1527
209. Schmaltz RM, Hanson SR, Wong C-H (2011) Chem Rev 111:4259–4307
210. Qian X, Sujino K, Palcic MM, Ratcliffe RM (2002) J Carbohydr Chem 21:911–942
211. Wong C-H, Haynie SL, Whitesides GM (1982) J Org Chem 47:5416–5418
212. Dumas DP, Ichikawa Y, Wong C-H, Lowe JB, Nair RP (1991) Bioorg Med Chem Lett 1:425–428
213. Sabesan S, Paulson JC (1986) J Am Chem Soc 108:2068–2080
214. Unverzagt C, Kunz H, Paulson J (1990) J Am Chem Soc 01:9308–9309
215. Sugiarto G, Lau K, Qu J, Li Y, Lim S, Mu S, Ames JB, Fishe AJ, Chen X (2012) ACS Chem Biol 7:1232–1240
216. Aharoni A, Thieme K, Chiu CPC, Buchini S, Lairson LL, Chen H, Strynadka NCJ, Wakarchuk WW, Withers SG (2006) Nat Methods 3:609–614
217. Farkas E, Thiem J (1999) Eur J Org Chem 1999:3073–3077
218. Fernández-González M, Boutureira O, Bernardes GJL, Chalker JM, Young MA, Errey JC, Davis BG (2010) Chem Sci 1:709–715
219. Rakić B, Withers SG (2009) Aust J Chem 62:510–520
220. Armstrong Z, Withers SG (2013) Biopolymers 99:666–674
221. Mackenzie LF, Wang Q, Warren RAJ, Withers SG (1998) J Am Chem Soc 7863:5583–5584
222. Kwan DH, Withers SG (2011) J Carbohydr Chem 30:181–205
223. Williams SJ, Withers SG (2000) Carbohydr Res 327:27–46

Part I
Analysing Protein–Carbohydrate Interactions at the Molecular Level

Chapter 2
Protein–Carbohydrate Interactions in Protein X-Ray Crystal Structures

This part of the results and discussion describes my work investigating the nature of PCIs. This primarily consists of the analysis of protein X-ray crystal structures containing carbohydrates. I then correlated the distributions of amino acids around carbohydrates to models of carbohydrate properties to identify trends. Finally, it describes studies to confirm the findings about carbohydrate–aromatic interactions by NMR spectroscopy. This first chapter describes the techniques that I used to analyse data from the protein structures, and the development of these tools.

Acknowledgements: The computer scripts for spatial analysis of amino acids proximal to carbohydrates were initially developed in conjunction with Dr. Gail Bartlett. The validation of the quality of raw data was carried out in collaboration with Dr. Jon Agirre (University of York).

2.1 Identifying Carbohydrates in Protein Structures

The RCSB PDB [1, 2] was used as a source of captured PCIs. The analysis of these interactions first required the identification and assignment of carbohydrate residues in the protein X-ray crystal structures in the database. The complexity inherent in carbohydrate structures (Sect, 1.1.1) made this non-trivial. This was complicated further as, while there are standardised residue labels for different carbohydrates in structures recorded in PDB format, these are inconsistent and not always used correctly [3]. There are labels for individual monosaccharides, but these do not always differentiate between anomers; some substituted monosaccharide units have separate labels, while for others the parent monosaccharide is used; and in some cases entire oligosaccharides have assigned labels that can be used in place of separate ones for individual monosaccharides.

K. L. Hudson, *Carbohydrate-Based Interactions at the Molecular and the Cellular Level*, Springer Theses, https://doi.org/10.1007/978-3-319-77706-1_2

Treatment of the connectivity in oligosaccharides and glycans is also inconsistent in structures from the PDB. It is more complex than for proteins due to the possibility of branching and also the presence of multiple examples of the same oligosaccharide. Within some structures carbohydrate ligands are labelled with the chain of the protein with which they interact, in others as separate individual chains, and sometimes all carbohydrates are given a single separate chain identifier. The oligosaccharide connectivity, i.e., the types of linkage between the different monosaccharides, is also not explicit within the PDB files, only implicit through the atomic coordinates.

The identity, substitution, and connectivity of carbohydrates determines how they interact with other species. To interpret fully the information about PCIs contained within the PDB all of these issues must be overcome.

2.1.1 Using GlyVicinity to Determine Carbohydrate Identity

The computer programmes contained within the CSS [3] address many of the pitfalls associated with identifying carbohydrates in protein structures of the PDB. There are features to determine the class and type (α/β, D/L) of monosaccharide residues, and also the substitution and connectivity within oligosaccharides and to amino acids. Outputs are given as LINUCS notation [4], which is used in this chapter to describe oligosaccharides concisely.[1]

GlyVicinity [5, 6] builds upon the CSS and, in addition to identifying carbohydrates, also determines the nearby amino acids within a user-determined threshold distance. This amino acid–carbohydrate distance is calculated between the closest two atoms of the two species. Other parameters that can be set include maximum X-ray crystal structure resolution and whether to include monosaccharide residues from non-covalently bound ligands, covalently attached glycans, or both. Results are obtained for specific monosaccharide species or oligosaccharides in the database.

The output from GlyVicinity gives the data for the monosaccharide residues identified in each carbohydrate-containing PDB file. This is separated by oligosaccharide chain, with the identity of each monosaccharide listed with all of the interacting amino acids and minimum inter-species distances. The HTML data includes the residue labels for the atoms of the carbohydrate and the amino acids for each atom pair; cross-reference of these labels to the coordinates in the original crystal structures deposited in the PDB allows investigation of the full species.

2.1.2 Parsing the GlyVicinity Output

To provide the data used for analysis in the study, an output was generated for all carbohydrate species identified by GlyVicinity. To give a large set of data to work

[1]See Sect. 1.1.1 for an explanation of LINUCS notation.

with, this was set at a maximum crystal structure resolution of 2.5 Å, for all amino acids with any atom within 4.0 Å of each carbohydrate residue, and for both ligands and glycans. These parameters were then refined in subsequent processing of the data.

Analysis of Oligosaccharides in High-Resolution Structures

As a first attempt to verify the feasibility of using the PDB to study PCIs, a script was developed that parsed the GlyVicinity HTML output and separated and grouped all of the different oligosaccharide chains represented. This was carried out for only crystal structures with a resolution of $\leq$1.6 Å, giving a total of 495 chains of 214 different types. This included 75 examples of *N*-linked GlcNAc disaccharide ([][ASN]{[(4+1)][b-D-GlcpNAc]{[(4+1)][b-D-GlcpNAc]{}}}, Table 2.1), while 148 oligosaccharides occurred only once in the data set.

The proximal amino acids for the different oligosaccharides were compared to investigate variations in interactions with proteins. Three oligosaccharides were chosen (Table 2.1, Fig. 2.1): cellobiose (CelB) ([][b-D-Glcp]{[(4+1)][b-D-Glcp]{}}, **18**, 12 examples), trehalose ([][a-D-Glcp]{[(1+1)][a-D-Glcp]{}}, **19**, 11 examples), and lactose (Lac) ([][b-D-Glcp]{[(4+1)][b-D-Galp]{}}, **20**, 10 examples). These provided informative comparisons, differing only in linkage (CelB vs. trehalose) or hydroxyl configuration at one position (CelB vs. Lac). These subtle variations served as a good test for the use of the PDB to elucidate carbohydrate binding—changes in the binding amino acids should show how these differences are discriminated by proteins.

To facilitate analysis and visualisation of the data contained within the GlyVicinity output, small working files containing only the coordinates of the carbohydrate and the detected proximal amino acids were created for each separate residue within the oligosaccharides. Investigating these files for the three different disaccharides revealed clear differences, especially conspicuous in the positioning of aromatic amino acids. For example, for the β-D-Glc residues of CelB, aromatic rings were frequently observed both above and below the plane of the carbohydrate ring, such that the faces of the aromatic ring were presented to the carbohydrate (Fig. 2.2a).

Table 2.1 Oligosaccharides with $\geq$10 examples, in the GlyVicinity output from structures with resolution $\leq$1.6 Å

Number	Oligosaccharide (LINUCS notation [4])
75	[][ASN]{[(4+1)][b-D-GlcpNAc]{[(4+1)][b-D-GlcpNAc]{}}}
19	[][b-D-Fruf]{[(2+1)][a-D-Glcp]{}}
14	[][ASN]{[(4+1)][b-D-GlcpNAc]{[(4+1)][b-D-GlcpNAc] {[(4+1)][b-D-Manp]{}}}}
12	[][<C12>]{[(1+1)][b-D-Glcp]{[(4+1)][a-D-Glcp]{}}}
12	[][b-D-Glcp]{[(4+1)][b-D-Glcp]{}}
11	[][a-D-Glcp]{[(1+1)][a-D-Glcp]{}}
10	[][b-D-Glcp]{[(4+1)][b-D-Galp]{}}

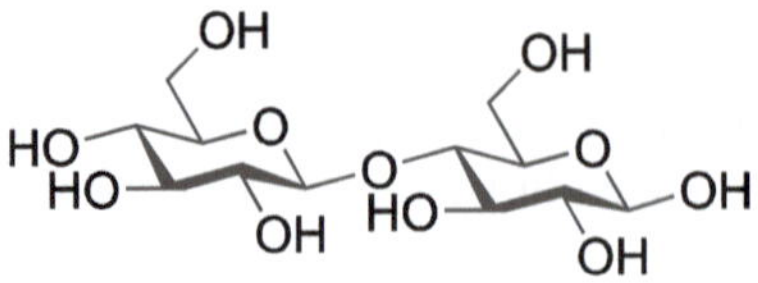

18

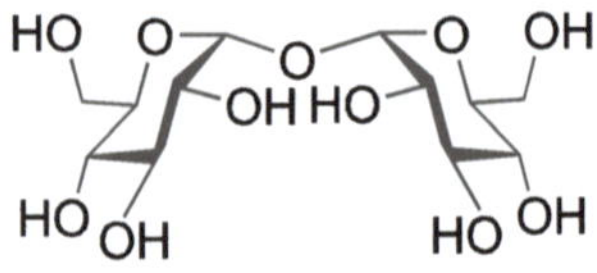

19

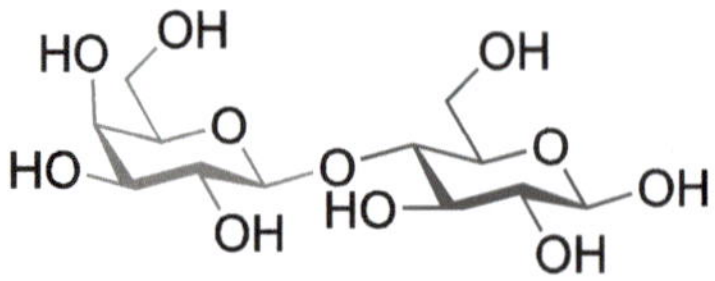

20

Fig. 2.1 Structures of disaccharides CelB (**18**), trehalose (**19**), and Lac (**20**)

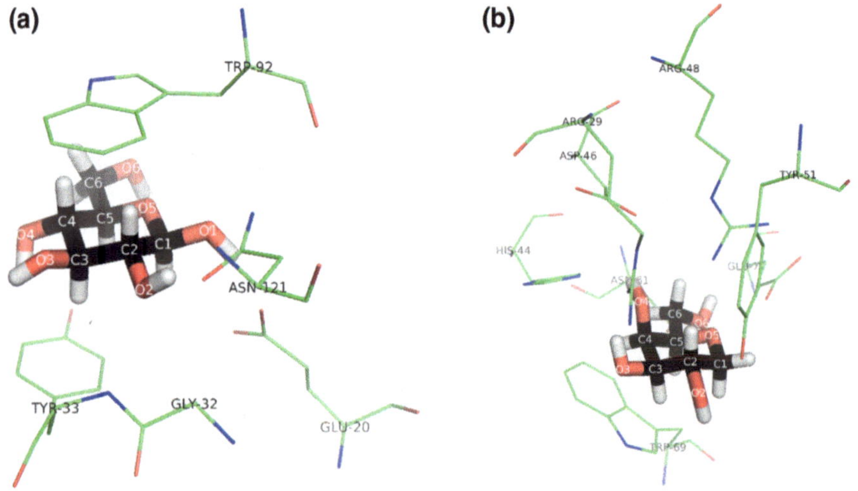

Fig. 2.2 Examples of monosaccharide residues with proximal amino acids obtained from first processing of PDB. **a** Typical β-D-Glcp residue from CelB (BGC 1136B from PDB ID 1UYY). **b** Typical β-D-Galp residue from Lac (GAL 401B from PDB ID 1WLD). Images generated with PyMOL [7]

In contrast to the β-D-Glc in CelB, the most common orientation of aromatics around the β-D-Gal in Lac were such that the plane of the aromatic ring was offset to the carbohydrate, presenting the face between C3, C4, and C5 (Fig. 2.2b). This is interesting as the offset mode of interaction is possible for β-D-Gal, where C(4)–OH is axial, but not for β-D-Glc, where the aromatic ring would sterically clash with the equatorial C(4)–OH. This suggests a mechanism for discrimination between the disaccharides. For trehalose, aromatic residues were on the whole less-frequently observed in proximity, with very few presenting faces to the carbohydrate.

Although these results were promising and demonstrated the potential of this approach, they also highlighted challenges for the oligosaccharide-focussed approach. The first of these is sample size: only eight oligosaccharides were represented ten or more times, and only one of these more than twenty times. This number of examples is low to determine general principles. Inconsistent labelling was also problematic: all three of the sample disaccharides were represented in some crystal structures as single residues, and in others with separate labels for the monosaccharides. A third problem was redundancy in the dataset—for one of the better represented oligosaccharides, cyclohexyl-hexyl-β-D-maltoside ([][<C12>]{[(1+1)][b-D-Glcp]{[(4+1)][a-D-Glcp]{}}}), all 12 examples were proximal to the same three amino acid residues in identical positions and orientations, despite all originating from separate PDB entries.

Analysis by Monosaccharide for Increased Sample Size

To address the issue of sample size, the same analysis was carried out on the entire GlyVicinity output without the maximum resolution threshold, i.e., using all of the data up to a resolution of 2.5 Å. To remove bias due to degeneracy in the data set, the protein structures were culled at maximum mutual sequence identity of 40% using the programme PISCES [8].

Even with the removal of structures with sequence identity, the higher resolution cut-off gave around twice the number of oligosaccharide examples as the previous analysis, at 980 in total, with 357 unique structures. The most common was a single β-*N*-linked GlcNAc ([][ASN]{[(4+1)][b-D-GlcpNAc]{}}). Also common were the *N*-linked GlcNAc disaccharide observed previously, free β-D-Glc ([][b-D-Glcp]{}), and octyl glucoside ([][octyl]{[(1+1)][b-D-Glcp]{}}), which is commonly used as a detergent in protein crystallisation [9] (Table 2.2). Again, there was high diversity in the oligosaccharide structures, with only eleven represented ten or more times.

Given the low occurrences of most of the individual oligosaccharides, I carried out the analysis of carbohydrates on more simple terms. I looked at differences between individual monosaccharide residues regardless of position within the oligosaccharide structures. This disregarded possible differences due to substitution, such as those already seen for the Glc residues in CelB, Lac, and trehalose, but was necessary for adequate sample sizes. Differences in modes of interaction for different monosaccharides would still allow elucidation of the mechanisms by which carbohydrates are bound by proteins. Carbohydrates provide a suitable model for wider ligand binding as there are many classes differentiated by only minor factors, such as changes in configuration at a single stereocentre, or more major ones, such as the presence or absence of functional groups that are charged at physiological pH.

The coordinates of the monosaccharides of interest were extracted from the oligosaccharides along with their proximal amino acids as identified by GlyVicinity. Overlaying the resulting sets of coordinates for a particular monosaccharide on the carbohydrate rings allowed visualisation of the 'coordination sphere' of all amino acids close enough to interact. This approach yielded a large amount of information, even for the less-well represented monosaccharides in the data set, such as β-D-Gal (Fig. 2.3).

Table 2.2 Oligosaccharides with ≥10 examples, in the sequence-culled data set from crystal structures with resolution ≤2.5 Å

Number	Oligosaccharide (LINUCS notation [4])
144	[][ASN]{[(4+1)][b-D-GlcpNAc]{}}
70	[][ASN]{[(4+1)][b-D-GlcpNAc]{[(4+1)][b-D-GlcpNAc]{}}}
38	[][octyl]{[(1+1)][b-D-Glcp]{}}
32	[][b-D-Glcp]{}
28	[][b-D-Fruf]{[(2+1)][a-D-Glcp]{}}
25	[][ASN]{[(4+1)][b-D-GlcpNAc]{[(4+1)][b-D-GlcpNAc] {[(4+1)][b-D-Manp]{}}}}
22	[][ASN]{[(4+1)][a-D-GlcpNAc]{}}
20	[][a-D-Glcp]{}
11	[][a-D-Glcp]{[(4+1)][a-D-Glcp]{}}
11	[][b-D-GlcpNAc]{}
10	[][a-D-Glcp]{[(1+1)][a-D-Galp]{}}

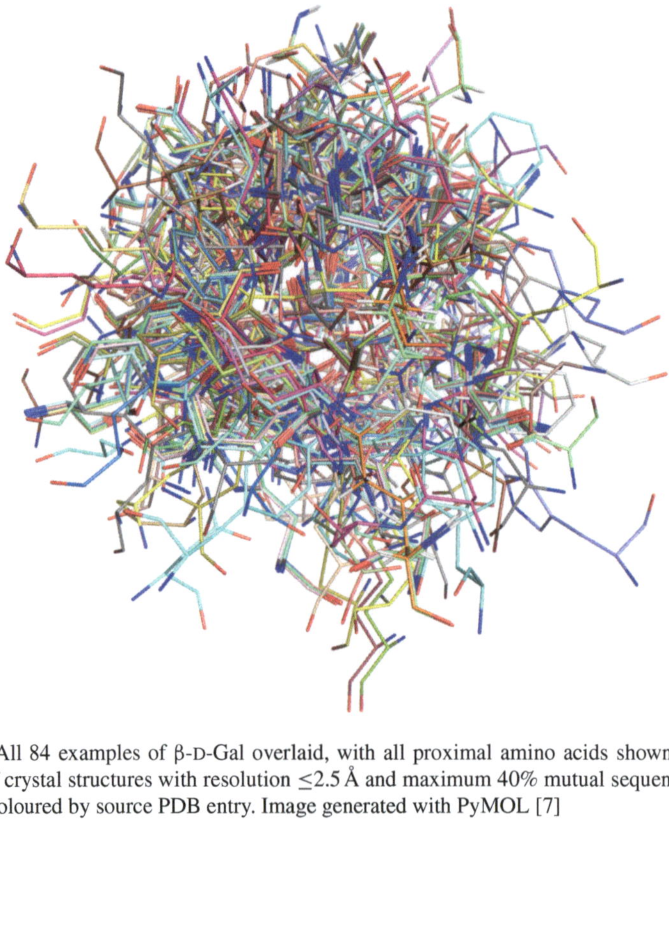

Fig. 2.3 All 84 examples of β-D-Gal overlaid, with all proximal amino acids shown. From the data set of crystal structures with resolution ≤2.5 Å and maximum 40% mutual sequence identity. Carbons coloured by source PDB entry. Image generated with PyMOL [7]

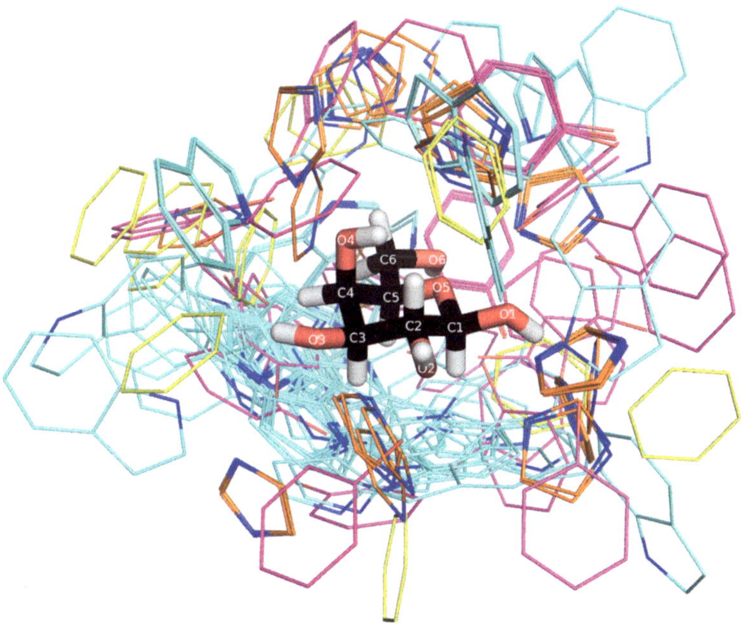

Fig. 2.4 All 84 examples of β-D-Gal overlaid, with aromatic moieties from proximal amino acids shown. From the data set of crystal structures with resolution ≤2.5 Å and maximum 40% mutual sequence identity. Typical monosaccharide residue shown. Carbon atoms coloured by residue: black = carbohydrate; cyan = Trp; pink = Tyr; yellow = Phe; orange = His. Image generated with PyMOL [7]

While considering the data in its entirety made drawing conclusions difficult, trends became apparent by focussing on specific amino acids. For example, clear biases were seen in the distribution and orientation of the four aromatic amino acids (histidine (His), phenylalanine (Phe), Trp, Tyr) around β-D-Gal (Fig. 2.4). This was particularly apparent for the Trp residues. Other monosaccharides present in comparable numbers, such as α-D-Glc and β-D-Man, also showed biased and distinct amino acid distributions.

2.1.3 Generating the Data Set of Carbohydrates with Proximal Amino Acids

These initial studies successfully established the utility of the PDB as a database of PCIs, and demonstrated differences in the interactions of different carbohydrates. A script was developed to convert the HTML data output from GlyVicinity into a format more readily usable for investigating PCIs in three dimensions. This generated a database of the working files containing the coordinates of all of the identified carbohydrates along with the proximal amino acids.

The script parsed the GlyVicinity output and cross-referenced to the original crystal structures in the PDB to obtain the appropriate coordinates. It also presented an opportunity to correct some errors within the GlyVicinity output, as well as to address redundancy in the data set. Errors that were corrected included removing duplicate entries from the GlyVicinity output, as well as identifying cases where the residue label corresponding to the identified carbohydrate was incorrectly assigned and attempting to find the correct residue using the recorded atom labels. Examples where entire oligosaccharides were given a single label in the crystal structure were discounted, to avoid issues determining the monosaccharide referred to by GlyVicinity.

In some cases certain pairs or groups of identical amino acids occupied almost the same positions relative to the carbohydrate in the files generated for separate monosaccharide residues, even after the mutual sequence identity cull. Most of these examples were due to duplicated binding sites in oligomeric crystal structures. These duplications were removed to prevent bias in the data set by identifying the composition of the proximal amino acids and discounting cases where they were identical. In some cases binding sites from different proteins were almost identical despite them being below the threshold for mutual sequence identity. This is possibly due to the presence of the same conserved carbohydrate binding module domains in diverse proteins [10], or convergent evolution to identical solutions for optimal carbohydrate-binding motifs [11]. These sets of similar examples were kept in the data set, as whether due to convergence or retention through evolution they must represent favourable and robust binding modes.

A script was developed to allow the easy investigation of specific monosaccharides using the labels assigned to monosaccharide residues in the GlyVicinity output to search the database of coordinate files for those that were relevant. As this relied upon GlyVicinity for identification of the residues, some false assignments were carried through, but these were straightforwardly removed manually upon visual inspection of the overlaid output files.

2.2 Analysis of the Composition and Distribution of Amino Acids Surrounding Carbohydrates

To help interpret the spatial distribution of amino acids around the carbohydrate residues, and also the visualisation of the data, a script was written to process the working PDB files that had been generated. This focussed only on the side chains of the amino acids, where the different functional groups are located. To find preferences for particular amino acids interacting with specific regions of the carbohydrate residues, the environment around the monosaccharide was divided into 14 'sectors' (seven each on the α- and β-faces, Fig. 2.5), and the relative populations of each

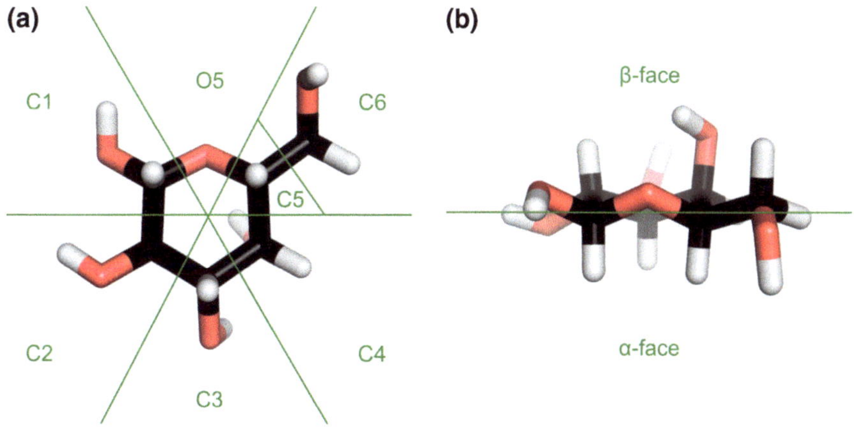

Fig. 2.5 Sectors into which centres of interacting amino side chains were divided, defined by nearest of the carbohydrate carbons or ring oxygen, as well as position relative to the ring. Shown for β-D-Gal, from **a** α-face and **b** side

amino acid in these sectors were determined. A 'centre' was assigned to the side chains (Table 7.1) and the nearest of the carbohydrate carbon atoms and ring oxygen found as well as the position relative to a plane through the carbohydrate ring. Recording the distance between this side chain centre and the carbohydrate atom also allowed for the application of distance thresholds if required in subsequent analyses. For example, amino acids where a main chain atom falls within the GlyVicinity cut-off but the side chain itself does not interact with the monosaccharide could be discounted.

For each carbohydrate residue in the data set, a new file was generated with dummy atoms placed at the amino acid side-chain centres, rather than the full coordinates of each amino acid. This allowed for easy visualisation of carbohydrate interactions (Fig. 2.6). To simplify interpretation of the data further and to investigate overall trends in preferred interactions the amino acids were separated in to three classes: those with aromatic (His, Phe, Trp, Tyr), polar (Asp, Asn, Arg, glutamine (Gln), Glu, Lys, serine (Ser), threonine (Thr)), and aliphatic (Ala, cysteine (Cys), Gly, Ile, Leu, methionine (Met), proline (Pro), valine (Val)) side chains.

Although the placement of amino acids into sectors greatly reduced the spatial detail in the data set, factors such as resolution of the diffraction data and conformational freedom of amino acid side chains in solution meant that interpreting at any greater detail was not necessarily valid.

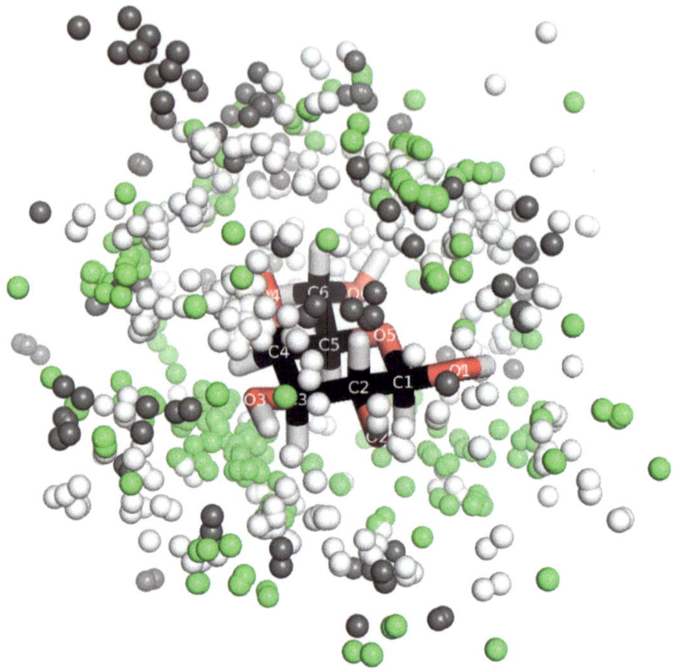

Fig. 2.6 All examples from the data set of β-D-Gal from ligands overlaid, with centres of all proximal amino acid side chains displayed as spheres. Amino acid colour code: aromatic = green; aliphatic = grey; polar = white. From the data set of crystal structures with resolution ≤2.5 Å and maximum 40% mutual sequence identity. Image generated with PyMOL [7]

2.3 Identifying CH–π Interactions

Carbohydrate–aromatic interactions are known to play an important role in binding of carbohydrates by proteins [12], particularly through the formation of CH–π interactions (Sect. 1.2.4) [13]. I chose the involvement of carbohydrates in CH–π interactions as a particular focus of this study, with the preliminary studies showing apparent differences in preferred orientations of aromatic amino acid side chains around different monosaccharides in the data set. Carbohydrates provide a particularly suitable model for studying the role of CH–π interactions in the wider context of ligand binding as the fixed chair conformation allows accurate prediction of the position of the C–H proton. Protein X-ray crystal structures are rarely of sufficient resolution to show protons, and so placing them precisely is difficult for more flexible ligands.

I used parameters adapted from those used by Brandl et al. in a study of interactions within proteins [14] to identify CH–π interactions in the data set (Fig. 2.7a). For this analysis, the parameters were defined in terms of the carbons bearing C–H bonds, given the aforementioned absence of density in the diffraction data for the C–H

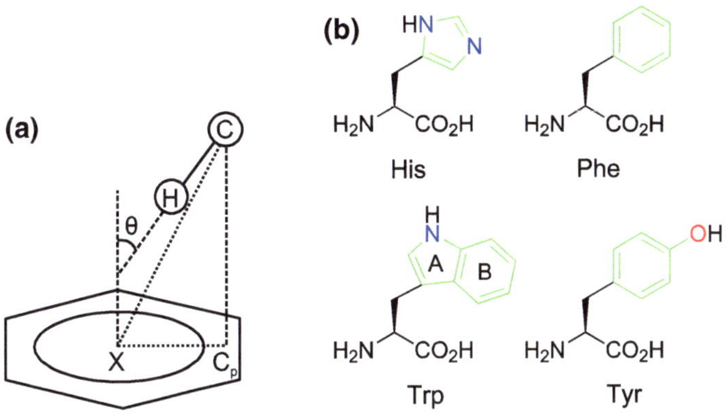

Fig. 2.7 **a** Parameters used to identify CH–π interactions [14]: CH–π angle (θ, ≤40°), CH–π distance (C–X, ≤4.5 Å), C-projection distance (C_p–X, ≤1.6 Å for His and TrpA; ≤2.0 Å for Phe, TrpB, Tyr). Reproduced from Ref. [15]. **b** Structure of proteinogenic aromatic amino acids with rings used to detect CH–π acceptors highlighted in green. For Trp, the five- and six-membered rings were treated separately, as TrpA and TrpB, respectively

protons in most of the data set. The CH–π acceptors were defined by the aromatic rings of the amino-acid side chains, with the indole moiety of Trp treated as two separate rings (Fig. 2.7b).

The first parameter defined was the distance between the carbon atom and the centre of the aromatic ring (C–X in Fig. 2.7a). For the CH–π interaction to contribute to carbohydrate binding the C–H proton and aromatic π-system must be proximal, within or close to van der Waals contact to allow orbital electrostatic interaction. The average distance from the C–H proton to the π-system in CH–π interactions is around 2.6–2.7 Å [13]. Including the typical C–H bond length of 1.09 Å for measurement to the carbon, and accounting for resolution and off-centre binding, the cut-off for CH–π interactions was set at 4.5 Å. Most interactions found in the study of proteins in the PDB by Brandl et al. had C–X distances of 3.5–4.0 Å [14].

CH–π interactions also require orientation of the aromatic ring such that the interaction of the π-system with the σ* C–H proton is overall attractive. Determining this for the aromatic amino acids in the data set was initially achieved by setting a 'tolerance' distance, as a maximum difference for the distance to the carbohydrate carbon for the nearest and farthest aromatic ring atoms. Several values for this distance were tested, scaled for ring size, but these gave unsatisfactory results that excluded rings perpendicular to but offset from the C–H bond (Fig. 2.8a). A more-appropriate parameter was the angle between the normal to the plane of the aromatic ring and the C–H bond (θ in Fig. 2.7a). For axial C–H bonds this is equivalent to the angle between the planes of the aromatic and carbohydrate rings. The correct angle for equatorial C–H bonds was calculated from the angle between the planes: the equatorial bond is 19.5° from the plane of the ring due to the fixed chair conformation of the carbohydrate rings and the 109.5° angle between bonds for sp^3-hybridized atoms. I used a maximum cut-off of 40° for this angle. This parameter differed slightly from

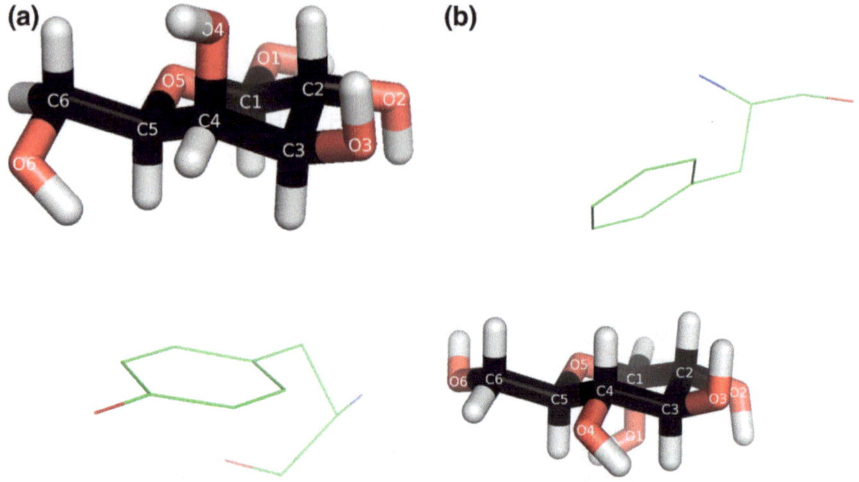

Fig. 2.8 Examples of errors from early methods of CH–π interaction detection. **a** Example of offset CH–π interaction missed by using 'tolerance' parameters rather than angles between planes of rings (GAL 500A and TYR 201A from PDB ID 1SO0). **b** Example of incorrectly identified CH–π interaction excluded by addition of C-projection parameter (GLC 1203A and PHE 175A from PDB ID 1V2B). Images generated with PyMOL [7]

the published study of protein structures, which uses the C–H–X angle [14], as I did not model the C–H proton. This calculation was only valid for carbohydrate rings in the 6-membered pyranose form, and so 5-membered furanoses (as identified by GlyVicinity) were not included in the study at this time.

The final parameter was the C-projection distance (C_p–X in Fig. 2.7a). This removed false-positive detections of interactions where the distance was within the limit and the angle to the C–H bond correct, but the π-system of the aromatic ring faces away from the C–H proton (Fig. 2.8b). It was measured as the distance between the C–H carbon projected onto the plance of the ring and the centre of the ring. The maximum cut-off for this depended on the size of the aromatic ring, such that interactions were only detected where the C–H proton was projected within the aromatic ring (Fig. 2.7a).

Where multiple carbohydrate C–H protons fell within the parameters such that they were classified as participating in a CH–π interaction with the same aromatic ring, the C-projection distance was used to assign the 'primary' interacting C–H bond. While each of these protons contribute to the carbohydrate–aromatic interaction, one C–H bond was chosen to facilitate data interpretation.

2.4 Validating the Quality of the Structural Data

When relying on data from a public repository, such as the PDB, it is important to account for errors that they may contain. The coordinates in PDB files are interpretations of electron density obtained from X-ray diffraction experiments. In many cases, these data are clear and the represented atomic coordinates unambiguous, but sometimes accurate assignment is more difficult. Errors in assignment are a particular problem for carbohydrates in the PDB, with many examples being given the incorrect conformation, orientation, anomer, or identity, or not being represented by adequate density at all [16]. This could be for several reasons, including flexibility in carbohydrates weakly bound by proteins, or poor chemical knowledge of carbohydrates by those interpreting the data.

Validation Using Privateer

The programme Privateer analyses carbohydrate residues in X-ray crystal structures for these issues by comparison to the original diffraction data [17]. It checks that the assigned coordinates are representative of the density, and also if they are reasonable for the carbohydrate in question, i.e., that the carbohydrate is in a low-energy conformation. Agreement to the diffraction data is quantified as a real space correlation coefficient (RSCC), with values ranging from 0 for no supporting density to 1 for perfect agreement [18]. The software also determines the conformation of the carbohydrate by calculating the Cremer–Pople coordinates [19], which correspond to an assigned label, such as $^4C_1$. Comparison of these values, as well as the bond lengths, from the carbohydrate coordinates in the crystal structure to a calculated minimum energy conformation gives a diagnostic for the feasibility of the assigned structure. Privateer also gives other data, such as the mean density and mean B-factor of the carbohydrate residue.

As a measure of correlation between the coordinates published in the PDB and the raw data, the RSCC was used to validate the quality of the carbohydrates contained within the data set. This revealed that the non-validated data set contained some examples with very good RSCC values, as well as some where the coordinates were poorly supported by the observed electron density (Fig. 2.9).

The quality indicated by the RSCC is related to the resolution of the data; a residue with an RSCC of 0.9 from a structure at 1.6 Å resolution is of better quality than a residue with the same RSCC from a structure at 3.0 Å resolution. The average RSCCs of carbohydrate residues in crystal structures from the PDB also decline with increasing structure resolution (Fig. 2.10). Therefore, the maximum resolution cut-off for crystal structures included in the final study was reduced to 2.0 Å.

Generating a Validated Database

For structures of 2.0 Å resolution, an RSCC value of 0.8 represents a good basis for the assigned structure in electron density, and so this was the minimum cut-off that I applied for the study. The raw diffraction data must be published and accessible for validation of the residues within a crystal structure by Privateer, and so I only used those with structure factors available. Over 30% of the structures that I had used in

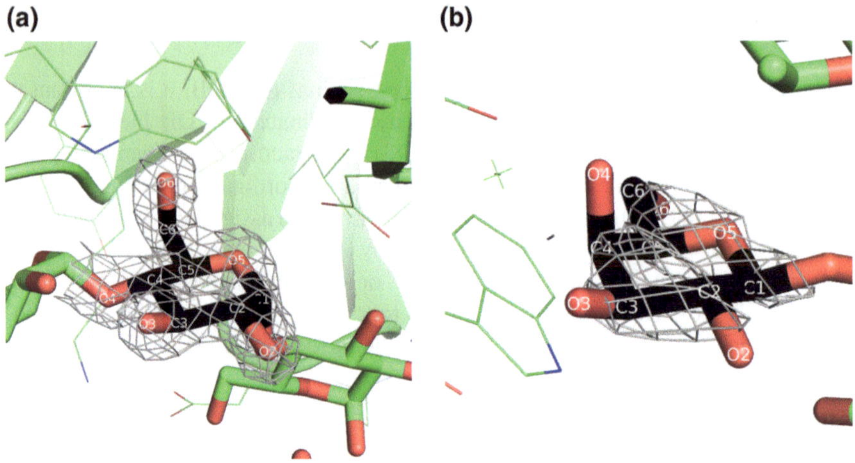

Fig. 2.9 Examples of correlations between X-ray diffraction data and assigned carbohydrate coordinates in structures in the PDB. **a** Good correlation (RSCC = 0.98, GLC 1104A from PDB ID 2C3W). **b** Poor correlation (RSCC = 0.51, GAL 1C from PDB ID 2OX9). Electron density shown as a mesh contoured at 2.0 σ. Images generated with PyMOL [7]

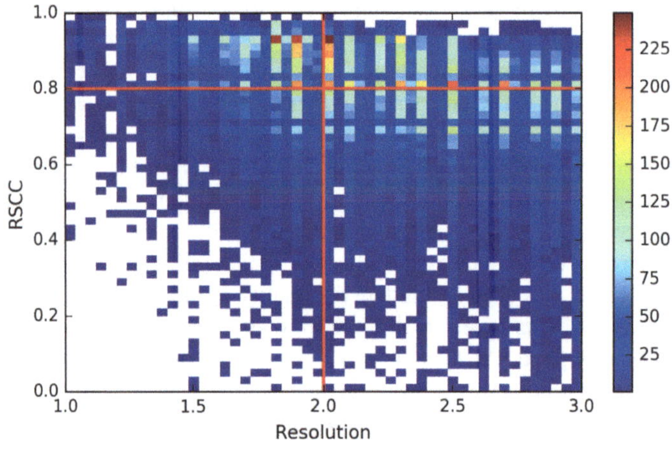

Fig. 2.10 2D histogram of RSCC values compared to the resolution for carbohydrate residues (with labels used in final study) in structures from the PDB with resolutions ≤3 Å. For the final study, only residues with an RSCC of ≥0.8 from structures of resolution ≤2.0 Å were used (both marked by red lines)

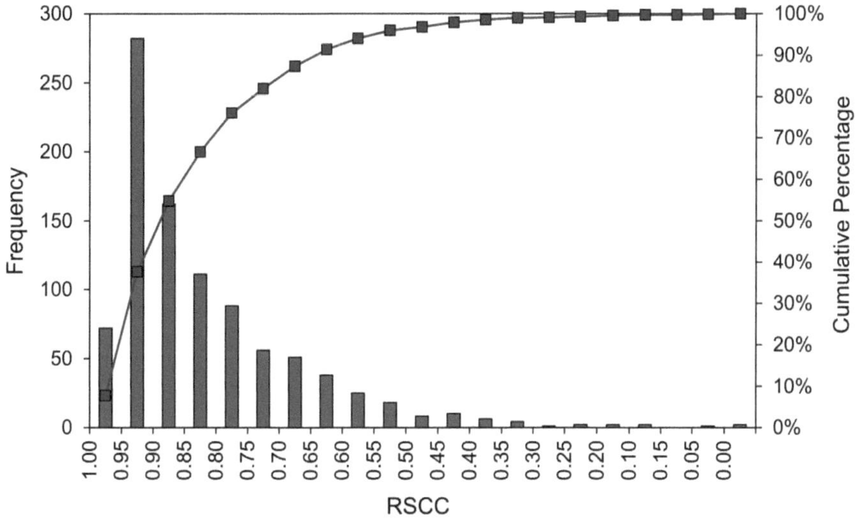

Fig. 2.11 Histogram and cumulative frequency of RSCC values for carbohydrate residues used before validation of PDB structures against diffraction data

the study prior to validation had an RSCC below the cut-off, and so did not represent reliable data (Fig. 2.11).

The determination of the monosaccharide conformation by Privateer was also useful for improving the quality of the data set. Both the calculation of CH–π angle and assignment of amino acid side chains to sectors around the carbohydrate assumed that the ring was in a chair conformation. Therefore, along with requiring an RSCC of ≥0.8 and resolution of ≤2.0 Å, only residues with a conformation of $^{4}C_{1}$ (for D-sugars) or $^{1}C_{4}$ (for L-sugars) were included in the final data set. By carrying out a Privateer analysis for all of the residue labels found in the non-validated data set across the entire PDB it was possible to create a new data set containing only valid carbohydrate residues.

At the point of regenerating the data set from only the validated monosaccharide residues in crystal structures, I also adapted the method for removing redundant structures. With the first method of culling maximum mutual sequence identity across all the carbohydrate-containing PDB structures, useful data where the same or similar proteins bound different carbohydrates was being lost. For example, two crystal structures of the same protein bound to Glc in one case and Gal in the other would contain separate information on how each is bound, but would be removed by the sequence-only cull. To overcome this, a separate cull was carried out for each class of monosaccharide residues, using the programme CD-HIT [20], with a maximum mutual sequence identity of 95% to remove identical proteins and point mutations.

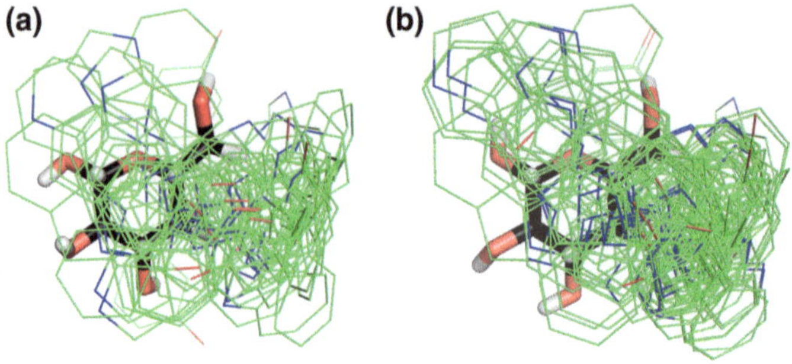

Fig. 2.12 All examples of β-D-Gal overlaid with CH–π accepting aromatic rings from amino acids, from **a** all PDB structures with resolution ≤2.5 Å, unvalidated, and **b** data validated such that RSCC ≥0.8 from PDB structures with resolution ≤2.0 Å. Images generated with PyMOL [7]

The addition of validation of the carbohydrate residues in the data set improved the quality of the data. Subtle changes were observed in the positioning of amino acids around the different monosaccharides in the validated data set compared to the non-validated, but the overall trends remained the same (Fig. 2.12).

2.5 Comparison of Data for Ligands and Glycans

With the database of validated carbohydrates with proximal amino acids in hand, I selected appropriate classes of monosaccharide for the detailed investigation of CH–π interactions. Carbohydrate–aromatic interactions are likely to be more important for the binding of neutral carbohydrates than for charged species; for the latter, electrostatic interactions with charged amino acid side chains will bind the carbohydrate strongly [21]. The seven neutral mammalian monosaccharides were chosen for analysis as both α- and β-anomers: Fuc, Gal, *N*-acetyl-D-galactosamine (GalNAc), Glc, GlcNAc, Man, and Xyl. These are also the most abundant neutral monosaccharide residues in the PDB [22]. Only the pyranose forms were considered, in order to be compatible with the developed scripts for identifying CH–π interactions, of which there were ≥10 examples for each monosaccharide residue.

The data set contained monosaccharide residues both from ligands (bound non-covalently to proteins) and glycans (bound covalently) in varying proportions for the different monosaccharides studied (Table 2.3). While for the unvalidated data set there were an approximately equal number of residues from glycans as from ligands, for the validated data two-thirds were from ligands, suggesting that the coordinates of carbohydrate glycans are more often based on questionable electron density. Many of the monosaccharides were observed mostly, or exclusively, from ligands, including both anomers of Gal and Glc.

The only monosaccharides where more than 20% of residues were from glycans were the two anomers of Man and β-D-GlcNAc, the latter of which was present

Table 2.3 Numbers of investigated monosaccharide residues present in data set, and percentage that were present as constituents of non-covalent ligands or glycans covalently linked to Asn, Ser, Thr, Glu, or Asp residues. Adapted from Hudson et al. [15]

Monosaccharide residue	Number	% Ligands	% Glycans				
			Asn	Ser	Thr	Glu	Asp
α-L-Fuc	79	84.8	15.2	0	0	0	0
β-L-Fuc	11	100	0	0	0	0	0
α-D-Gal	43	100	0	0	0	0	0
β-D-Gal	143	97.9	0.7	0	1.4	0	0
α-D-GalNAc	35	88.6	0	5.7	5.7	0	0
β-D-GalNAc	32	100	0	0	0	0	0
α-D-Glc	179	98.9	0	0	0	0.6	0.6
β-D-Glc	219	99.5	0	0	0	0	0.5
α-D-GlcNAc	25	96.0	4.0	0	0	0	0
β-D-GlcNAc	564	22.3	77.7	0	0	0	0
α-D-Man	159	57.9	19.5	17.6	5.0	0	0
β-D-Man	68	63.2	36.8	0	0	0	0
α-D-Xyl	36	100	0	0	0	0	0
β-D-Xyl	61	90.2	6.6	0	0	3.3	0
Total	1654	66.2	31.0	1.8	0.7	0.2	0.1

in *N*-glycans in 78% of cases. The prevalence of these three monosaccharides in glycans is not surprising, as natural *N*-linked glycans are always built upon a core of a β-D-GlcNAc linked to the Asn, followed by a further β-D-GlcNAc, and then Man residues [23]. These are then the components most closely associated with the protein, meaning they are the most likely to meet the criterion of being within 4 Å of an amino acid, and also be constrained enough to allow good density to be observed by X-ray diffraction.

Comparison of the four monosaccharides for which there were enough examples in the data set showed differences in the interaction with amino acids for residues from ligands compared to those from glycans (Table 2.4). In all cases there were on average fewer proximal amino acids for residues from glycans, around two thirds the number seen for examples from ligands. There was an even greater difference in the average participation in CH–π interactions, with less than a quarter of the number for glycans on average. Thus, non-covalent interactions are less important for glycans, as the binding of these is dominated by the much stronger covalent linkage.

Ligands non-covalently bound to proteins often represent known or putative binding partners of the proteins in the natural binding site. PCIs that are not representative of naturally occurring interactions, such as those involving detergents, e.g., octyl glucoside, still represent a minimum energy non-covalent interaction favoured during crystallisation. While the role of some covalently attached glycans is to modulate

Table 2.4 Comparison of number of proximal amino acids and participation in CH–π interactions for monosaccharide residues from ligands and from glycans, for which there were sufficient examples (≥10 of each). Adapted from Hudson et al. [15]

Monosaccharide anomer	L-Fuc	D-GlcNAc	D-Man		Total
	α	β	α	β	
Ligands					
Number	67	126	92	43	328
Proximal Amino Acids					
Total	416	757	523	190	1886
Average	6.21	6.01	5.68	4.42	5.75
CH–π Interactions					
Total	28	59	20	33	140
Average	0.42	0.47	0.22	0.77	0.43
Glycans					
Number	12	438	64	15	529
Proximal Amino Acids					
Total	39	1812	2224	33	2108
Average	3.25	4.14	3.50	2.20	3.98
CH–π Interactions					
Total	0	40	2	2	44
Average	0	0.09	0.03	0.13	0.08

protein structure through interaction with amino acid residues [24], for many others it is to bind other entities, such as lipids or other proteins. As these binding partners are generally not present in the protein crystals, the intramolecular interactions observed for glycans may not represent optimal PCIs, merely the best orientation possible within the constraints of the covalent linkage. Indeed, evolution may have led to minimisation of potential competing non-covalent interactions with the protein to which the glycans are attached for those with a role in binding other molecules.

Given the interest in elucidating non-covalent carbohydrate-based interactions, and the greater number and importance of these interactions for ligands over glycans, I carried out the full investigation and interpretation only for monosaccharide residues in the data set from carbohydrates bound non-covalently to proteins.

References

1. RCSB Protein Data Bank. http://www.rcsb.org/pdb
2. Berman HM, Westbrook J, Feng Z, Gilliland G, Bhat TN, Weissig H, Shindyalov IN, Bourne PE (2000) Nucleic Acids Res 28:235–242
3. Lütteke T, Frank M, von der Lieth C-W (2004) Carbohydr Res 339:1015–1020

4. Bohne-Lang A, Lang E, Förster T, von der Lieth C-W (2001) Carbohydr Res 336:1–11
5. Rojas M, GlyVicinity. http://glycosciences.de/tools/glyvicinity/
6. Lütteke T, Frank M, von der Lieth C-W (2005) Nucleic Acids Res 33:D242–D246
7. The PyMOL Molecular Graphics System, Version 1.7.0.3. 2014
8. Wang G, Dunbrack RL (2005) Nucleic Acids Res 33:W94–W98
9. Seddon AM, Curnow P, Booth PJ (2004) Biochim Biophys Acta 1666:105–117
10. Boraston AB, Bolam DN, Gilbert HJ, Davies GJ (2004) Biochem J 382:769–781
11. Taylor ME, Drickamer K (2014) Curr Opin Struct Biol 28C:14–22
12. Asensio JL, Ardá A, Cañada FJ, Jiménez-Barbero J (2013) Acc Chem Res 46:946–954
13. Nishio M (2011) Phys Chem Chem Phys 13:13873–13900
14. Brandl M, Weiss MS, Jabs A, Sühnel J, Hilgenfeld R (2001) J Mol Biol 307:357–377
15. Hudson KL, Bartlett GJ, Diehl RC, Agirre J, Gallagher T, Kiessling LL, Woolfson DN (2015) J Am Chem Soc 137:15152–15160
16. Agirre J, Davies G, Wilson K, Cowtan K (2015) Nat Chem Biol 11:303
17. Agirre J, Iglesias-Fernández J, Rovira C, Davies GJ, Wilson KS, Cowtan KD (2015) Nat Struct Mol Biol 22:833–834
18. Brown EN, Ramaswamy S (2007) Acta Crystallogr D63:941–950
19. Cremer D, Pople JA (1975) J Am Chem Soc 97:1354–1358
20. Li W, Godzik A (2006) Bioinformatics 22:1658–1659
21. Gandhi NS, Mancera RL (2008) Chem Biol Drug Des 72:455–482
22. Werz DB, Ranzinger R, Herget S, Adibekian A, von der Lieth C-W, Seeberger PH (2007) ACS Chem Biol 2:685–691
23. Kornfeld R, Kornfeld S (1985) Annu Rev Biochem 54:631–664
24. Shental-Bechor D, Levy Y (2009) Curr Opin Struct Biol 19:524–533

Chapter 3
The Nature of Protein–Carbohydrate Interactions

3.1 Analysis of Amino Acids Proximal to Carbohydrates in Protein Crystal Structures

The composition of the amino acids at the binding sites of carbohydrates was analysed to find the preferred side-chain functional groups for carbohydrate-based interactions. Deeper investigation of the spatial orientations of different classes of amino acids around the monosaccharide residues was then carried out to find favoured modes of interaction, and these were related to the properties of the interacting species.

3.1.1 The Composition of Amino Acids Around Carbohydrates

Some common trends were seen across all of the monosaccharides that were investigated. Generally, occurrence in carbohydrate binding sites decreased with increasing amino acid hydrophobicity (Fig. 3.1). This reflects the hydrophilic nature of carbohydrates, which present many water-like hydroxyl groups. However, there were some notable exceptions to the general trend. The aromatic amino acids were some of the best-represented hydrophobic residues, indeed, Tyr and Trp were both in the four most-common amino acids, within error.

The relative occurrences of the different amino acids are only instructive in context, after taking account of the natural occurrences of the different amino acids in proteins. It is important to eliminate other effects on amino acid prevalence, such as genetic preference, favoured amino acids on the surface of proteins, or amino acid surface area.

Normalising for Amino Acid Incidence

The occurrence of each amino acid proximal to carbohydrates was normalised by calculation of a propensity compared to another distribution. The propensity is the

K. L. Hudson, *Carbohydrate-Based Interactions at the Molecular and the Cellular Level*, Springer Theses, https://doi.org/10.1007/978-3-319-77706-1_3

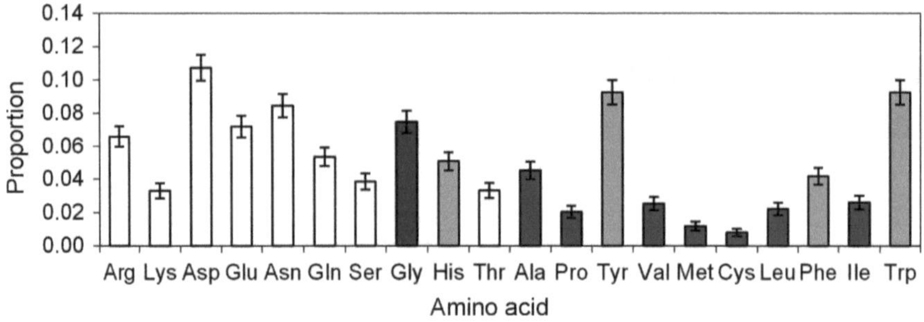

Fig. 3.1 Distribution of amino acids proximal to all investigated monosaccharide residues. Amino acids listed in order of increasing hydrophobicity [1]. Colour code: aromatic = green; aliphatic (non-aromatic hydrophobic) = grey; hydrogen-bonding = white. Error bars represent 95% confidence assuming a normal approximation of a binomial distribution

proportion of all amino acids proximal to carbohydrates divided by the proportion in the comparison data set, and so residues with propensity >1 are over-represented and <1 are under-represented in carbohydrate-binding sites. Propensities were calculated compared to occurrence in the UniProtKB/Swiss-Prot database of verified protein sequences [2, 3]. This database represents the relative occurrences of the amino acids in known proteins, and so the normalisation takes account of general preferences for the different amino acids in nature.

This showed stark differences in the propensity of different classes of amino acids near to carbohydrates (Fig. 3.2a). As with the raw proportions, propensities were generally lower for more-hydrophobic amino acids. Of the polar amino acids, those with hydrogen-bond-accepting side chains had the higher propensity in binding carbohydrates, particularly Asp and Asn, which each occurred about twice as often near carbohydrates as across proteins in general. Hydrogen bond donors Lys, Ser, and Thr were disfavoured. The non-aromatic hydrophobic amino acids also occurred less often, especially those with aliphatic side chains, with Ile, Leu, and Val less than half as common.

Aromatic amino acids are relatively rare in natural proteins, and so normalisation revealed an even more distinct preference for these near to carbohydrates. The only hydrophobic residues with propensities above one had aromatic side chains, and in fact Tyr and Trp were the amino acids with the highest propensities of all, at around three and nine, respectively. His had a propensity close to those of Asp and Asn, which is interesting as all three can act both as hydrogen-bond donors and acceptors. Phe was found near to carbohydrates at approximately the same rate at which it occurs across all protein structures, making it still much better represented than the non-aromatic residues of comparable hydrophobicity.

Alternative Methods of Normalisation

To eliminate other possible reasons for relative occurrences of proximal amino acids they were compared to expected distributions by several alternative methods, but the trends remained the same in all cases. First, it was considered that differences in

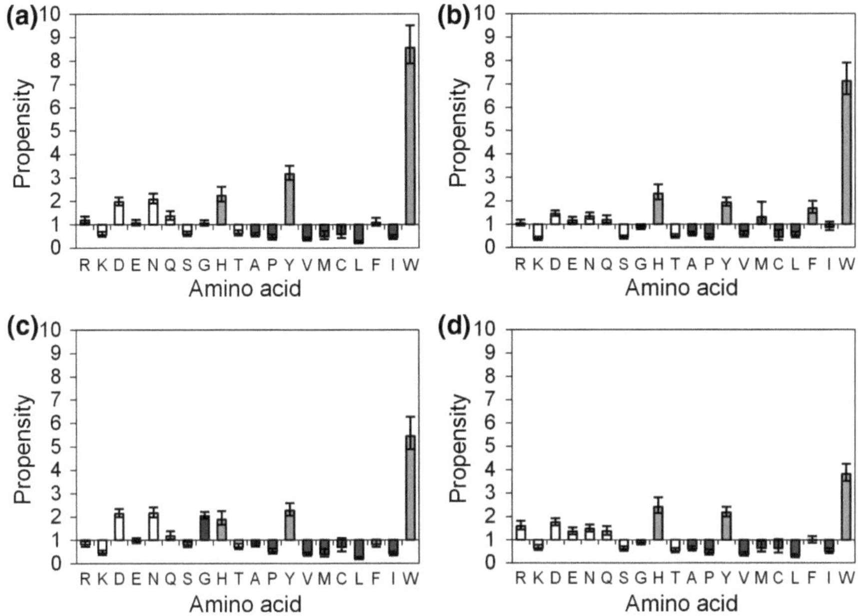

Fig. 3.2 Four different measures of propensity of amino acids proximal to carbohydrates. Adapted from Hudson et al. [20]. **a** Compared to occurrence in all proteins in UniProtKB/Swiss-Prot [2]; **b** Compared to amino acids on protein surfaces [4]; **c** Compared to representation of amino acids in UniProtKB/Swiss-Prot [2] corrected for total amino acid surface area [4]; **d** Compared to representation of amino acids in proteins from which the PCI database was derived. Amino acids listed in order of increasing hydrophobicity [1]. Colour code: aromatic = green; aliphatic (non-aromatic hydrophobic) = grey; polar = white. Propensity = (proportion of amino acids proximal to carbohydrates)/(proportion of amino acids in reference database). Error bars represent 95% confidence assuming a normal approximation of a binomial distribution

the composition of amino acids near carbohydrates might reflect the fact that most carbohydrate-binding sites are on the exterior protein surface. Calculating propensity compared to the composition of protein surface amino acids [5], however, made little difference to the rank and relative orders (Fig. 3.2b). Polar amino acids are favoured on solvent-accessible protein surfaces over hydrophobic ones, and the propensity of hydrogen-bonding residues was reduced by this normalisation compared to across protein structures as expected. The propensities of Tyr and Trp were also slightly reduced, but the aromatics were still the only hydrophobic residues occurring more frequently than expected near to carbohydrates.

The second possible factor that I tested was amino acid surface area, considering that residues that occupy more space may be more likely to be found adjacent to carbohydrates. Correcting the propensities compared to the UniProtKB/Swiss-Prot database for total amino acid surface area [5] made little difference to the propensities either (Fig. 3.2c). The propensity of Trp, as the amino acid with the largest surface area, did reduce to around six, but it was still the most over-represented residue proximal to carbohydrates.

Finally, I considered the possibility that the composition of the proteins from which the data set had been derived deviated from the composition across all protein sequences. Recalculating the propensities against the amino-acid occurrences across all of the proteins that I used led to a reduction in the propensity for Trp, to around four, but overall trends were the same as observed in the other cases (Fig. 3.2d). Thus, Trp is over-represented in carbohydrate-binding proteins in general, but nonetheless it remains favoured in carbohydrate binding sites. The polar amino acids were generally slightly favoured and the hydrophobic disfavoured, except for the aromatic residues. That the carbohydrate-containing protein structures used in this analysis had such an unusually high representation of Trp is itself interesting. This could be due to binding to further carbohydrate residues not included in the data set, or they could contribute to binding with carbohydrate through transient non-specific interactions.

The relative propensities of the different classes of amino acids proximal to carbohydrates are informative about how carbohydrates bind to proteins. The overall increased tendency for polar, hydrogen-bonding amino acids is not surprising, given the many hydroxyl groups presented by carbohydrates, however they are perhaps not as well-represented as would be expected. Meanwhile, the general absence of hydrophobic residues suggests that a hydrophobic effect does not play an important part in carbohydrate binding per se. The substantial over-representation of aromatic residues, particularly Tyr and Trp, indicates a specific role of carbohydrate–aromatic interactions in PCIs.

3.1.2 Amino Acid Density and Composition Around Different Monosaccharides

The almost 1100 examples of the seven chosen monosaccharide residues in the data set interacted on average with just under six amino acids each. The high occurrence of aromatic amino acids near to carbohydrates meant that there were more examples of these four residues than the eight non-aromatic hydrophobic amino acids combined, at 30 and 23% of all residues, respectively (Table 3.1).

Both the density and composition of nearby amino acids varied for the different monosaccharides. β-D-Man typically has the fewest nearby interacting amino acids of all of the studied examples, at under 4.5 on average. α-D-GlcNAc has the most nearby amino acids, at almost 7.5 on average, which may partly be due to the fact that as an *N*-acetyl aminosugar it is larger in terms of surface area and number of atoms than the simple hexoses. In five of the seven cases, the α-anomer interacted with more amino acids than the β-anomer, on average. In terms of composition, aromatic amino acids made up only 20% of amino acids near to α-D-Man, wherease they accounted for 41% of those around β-D-Man. The non-aromatic hydrophobic residues ranged from 15% of those nearby for α-D-Man to 29% of those proximal to

Table 3.1 Composition, facial distribution, and participation in CH–π interactions for amino acids proximal to the investigated monosaccharide residues from ligands in the data set. Adapted from Hudson et al. [20].

Monosaccharide Anomer	L-Fuc		D-Gal		D-GalNAc		D-Glc		D-GlcNAc		D-Man		D-Xyl		All
	α	β	α	β	α	β	α	β	α	β	α	β	α	β	
Number	67	11	43	140	31	32	177	218	24	126	92	43	36	55	1095
Proximal Amino Acids															
Total	416	75	292	752	215	171	1072	1287	179	757	523	190	160	298	6387
Average	6.21	6.82	6.79	5.37	6.94	5.34	6.06	5.90	7.46	6.01	5.68	4.42	4.44	5.42	5.83
Standard Deviation	3.02	0.98	2.77	3.01	2.43	2.04	3.10	3.15	3.13	2.87	2.84	2.77	2.18	2.62	
% Aromatic	29	36	23	32	24	26	30	33	28	29	20	41	30	37	30
% Aliphatic	22	29	22	17	29	25	26	22	26	24	24	15	20	19	23
% Polar	49	35	55	51	47	49	44	45	46	47	56	44	50	44	47
α-Face Amino Acids															
Average	0.7	1.1	1.1	1.3	2.0	1.0	0.8	1.2	1.5	1.1	0.7	0.8	0.9	1.6	1.1
% Aromatic	59	92	28	55	39	65	34	41	19	39	44	69	41	48	44
% Aliphatic	19	00	18	04	06	09	31	15	40	20	18	02	35	16	17
% Polar	22	08	54	41	55	26	35	44	41	41	38	29	24	36	39
β-Face Amino Acids															
Average	2.1	2.2	2.8	1.7	2.5	1.7	2.1	1.6	1.6	1.2	2.1	1.3	1.3	1.7	1.8
% Aromatic	20	04	25	15	21	04	40	25	39	23	08	06	31	26	23
% Aliphatic	20	17	19	18	11	26	20	24	11	24	32	18	19	16	21
% Polar	60	79	56	67	68	70	40	51	50	53	60	76	50	58	56
CH–π interactions															
Total	28	18	28	140	22	25	107	172	10	59	20	33	9	50	721
Average	0.42	1.64	0.65	1.00	0.71	0.78	0.60	0.79	0.42	0.47	0.22	0.77	0.24	0.91	0.66

α-D-GalNAc and α-L-Fuc. In all cases aromatic residues made up a larger proportion of the amino acids interacting with β-anomers than with the α-anomers, mostly with a corresponding reduction in the proportion of nearby aliphatic residues.

Starker differences were seen between the composition of amino-acid side chains interacting specifically with the α- and β-faces of the monosaccharide residues. These were identified as the amino acids for which the assigned side chain 'centre' was within 6 Å of any carbohydrate carbon or the ring oxygen, and the particular face determined by the position of this centre relative to a plane through the carbohydrate ring (Fig. 2.5, Sect. 2.2).

In general, there were fewer amino acids interacting with the β-faces of the monosaccharides, around one on average, than with the α-faces, approximately two on average. A greater proportion of amino acids that interacted with the β-face were aromatic than with the α-face for the same monosaccharides. This was particularly true for the β-anomers of monosaccharides with axial hydroxyl groups, such as β-D-Gal and β-D-Man, for which aliphatic side chains were almost completely absent. For these monosaccharides in particular, and all the cases to a lesser extent, a larger proportion of the amino acids were polar on the α-face.

These observations can be rationalised in terms of the carbohydrate structures. For β-D-Glc, with all equatorial hydroxyls, the primary difference between the α- and β-faces is that the former has three C–H groups, compared to two for the latter, so there are more C–H bonds to interact with non-polar amino acids on the α-face (Fig. 3.3). Adding to this, the chair conformation places the electron-rich ring oxygen closer to the β-face, and as the oxygen valence orbitals are sp^3-hybridised and so tetrahedral an orbital occupied by one of its lone pairs is pseudo-axial and projects towards that face. This oxygen can act as a hydrogen-bond acceptor, favouring interactions with a polar amino acid on the β-face.

These effects are more pronounced for the β-anomers with an axial hydroxyl, such as β-D-Gal (Fig. 3.3), which add a polar group to the β-face and removes a C–H compared to β-D-Glc. The opposite trend is observed when comparing α- to β-anomers, such as α-D-Glc, which exchange a C–H for an axial hydroxyl on the α-face. It is interesting that while additional C–H groups on a carbohydrate face lead to more

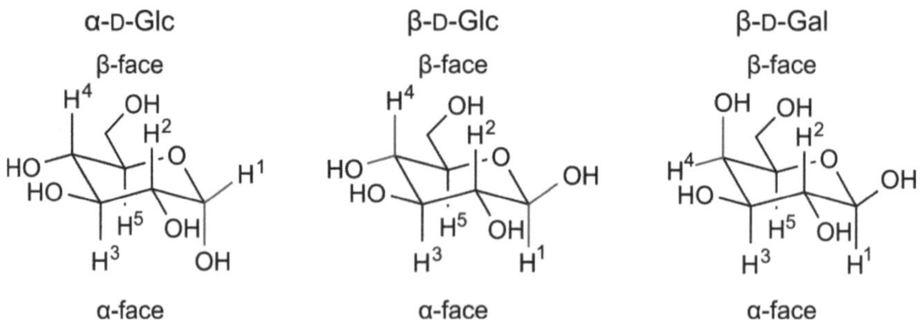

Fig. 3.3 Structure of α-D-Glc, β-D-Glc, and β-D-Gal

interacting aromatic amino acids, there is a corresponding reduction in the proportion of interacting aliphatic side chains. This again suggests that a general hydrophobic effect is not dominant, but rather a specific carbohydrate–aromatic interaction.

3.2 Distribution of Amino Acids Around Carbohydrates

Further information can be derived about how carbohydrates interact with proteins from the spatial distribution of amino acids around the monosaccharide residues. I focussed upon aromatic and non-aromatic hydrophobic residues to understand further the preference for aromatic residues. Given that these groups mostly interact with the carbohydrate C–H groups [6, 7], I analysed the composition of amino acid side chains interacting with the α- and β-faces where these bonds are found. The assigned side-chain 'centres' were used (Sect. 2.2), with a maximum distance from the carbohydrate carbons or O5 of 6 Å, as beyond this distance any interaction with the carbohydrate C–Hs would be minor or at least indirect. The aromatic and aliphatic side chain centres were visualised and assigned to a sector around the monosaccharide by finding the nearest C–H carbon (Fig. 3.4 and Sect. 7.2.2).

For β-D-Glc, with all its C–OH groups equatorial (Fig. 3.4a), the aromatic and aliphatic side chains interacting with each face tracked with the position of the C–H bonds, i.e., for a given carbon the greater proportion of aromatic and aliphatic amino acids was found on the face where that carbon's C–H is located (Fig. 3.4c). So the aromatic residues were located on the α-face nearest to C1, C3, and C5 where the C(1)–H, C(3)–H, and C(5)–H bonds are (Fig. 3.4a), while those nearest to C2 and C4 the aromatic residues were on the β-face. Around C6 an approximately equal proportion of the amino acids were aromatic and aliphatic on each face. This could be due to rotation around the C5–C6 bond, meaning that the C(6)–H bonds are not constrained to a particular face. Also, for the carbohydrate ω-angle (O5-C5-C6-O6 dihedral) favoured in solution and crystal structures [8, 9], one C(6)–H is on each face.

The aromatic amino acids also tracked with the C–H bonds for β-D-Gal. The only difference to β-D-Glc is the configuration at C4 (Fig. 3.4a), and for β-D-Gal the aromatic residues nearest to C4 are mostly found on the α-face. Indeed, the preference for aromatic residues proximal to the C–H bonds was more exaggerated for Gal, with C(4)–H and C(5)–H interacting almost exclusively with aromatic side chains, to the exclusion of aliphatics (Fig. 3.4b, c).

While β-D-Gal exchanges a β-face C–OH for an α-face C–H compared to β-D-Glc, α-D-Glc instead exchanges an α-face C–OH for a β-face C–H (Fig. 3.4a). This leads to an increased proportion of aromatic side chains near C–H bonds on the β-face for α-D-Glc compared to β-D-Glc, the opposite case to that for β-D-Gal.

The differences in the proportion of aromatic amino acids near to different carbohydrate C–H bonds suggest that not all of these groups interact in the same way, and this varies both between and within different monosaccharide residues. Some, such as the C(4)–H and C(5)–H of β-D-Gal, interact almost exclusively with aromatic

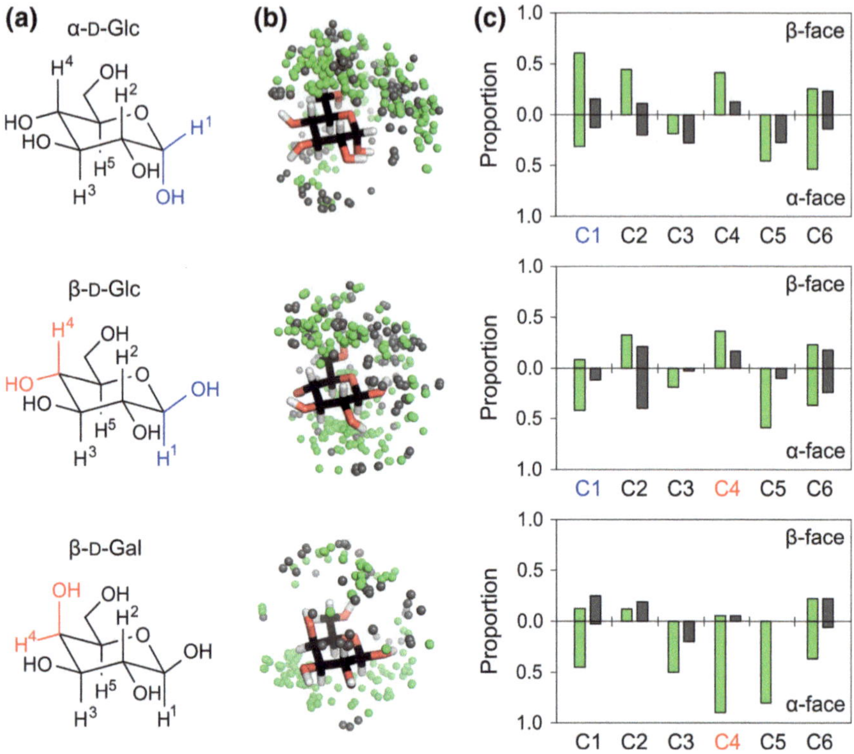

Fig. 3.4 Distribution of aromatic (green) and aliphatic (grey) amino acid side chains interacting with faces (i.e., within 6 Å of a carbohydrate carbon or O5) of α-D-Glc, β-D-Glc, and β-D-Gal. Adapted from Hudson et al. [20]. **a** Monosaccharide structures with C–H protons labelled systematically. **b** Spatial distribution of centres of side-chains, represented as spheres, around all examples of each monosaccharide from the data set overlaid. **c** Proportion of amino acid side chains nearest to each carbohydrate carbon that are aromatic and aliphatic on each face

residues, while for others they make up a much smaller proportion of nearby amino acids. These preferences do not correlate with the proportion of proximal aliphatic side chains. In fact, those C–H bonds with a higher preference for aromatics interact less often with aliphatics. This supports the importance of carbohydrate–aromatic interactions over simple hydrophobic effects.

3.3 Modelling the Properties of Carbohydrates and Aromatic Rings

Given the variations in the interactions with amino acids of different carbohydrate C–H bonds, and the indications that this was not determined by the hydrophobic effect, I investigated what properties were important in the carbohydrate–aromatic interactions. The key difference between aromatic moieties and the aliphatic side chains is the electronegative π-systems that can interact with the electropositive C–H protons in CH–π interactions [10]. The electrostatic component of this interaction would be greatest for species with the largest charge difference, and so if it were important in carbohydrate–aromatic interactions then more-electropositive C–H protons would interact preferentially with aromatic moieties. Therefore, I investigated the electronics of the different aromatic and monosaccharide residues and looked for any correlation to participation in carbohydrate–aromatic interactions.

Modelling the Electronic Properties of Molecules

The electrostatic surface potential ESP shows the three-dimensional charge distribution of a molecule [11, 12]. A surface is defined at a set percentage of total electron density of the molecule, and at each point on the surface the total charge contributed by each of the atoms in the molecule, scaled for their distance from the point, is calculated. This gives the potential electrostatic interaction with a charged species, and all of the points across the surface form the ESP. Therefore, the ESP provided a measure for the electronic properties of the carbohydrate and aromatic species in carbohydrate–aromatic interactions.

Ideally, ESPs for the monosaccharides would have been generated directly from representative examples taken directly from protein crystal structures in the data set. However, in the most part these did not contain coordinates for the protons as hydrogen atoms can only be determined from data at high resolution ($\leq$1.0 Å). Therefore, protons were added to a canonical structure, which was then optimised to a minimal energy conformation using density functional theory (DFT) calculations (B3LYP/6-31+(d)) in the computational chemistry programme Gaussian09 [13]. DFT is a modelling method based on quantum mechanical calculations of electron density functionals of molecules [14]. These minimised structures were checked to ensure that the adopted conformations were representative of those in the data set, and in all cases the minimised conformation was that which was most common in the crystallographic database, with the ω-angle most favoured in solution [9] (Fig. 3.5a).

Once the minimised conformations had been computed, ESPs were then generated from an energy calculation by the Hartree-Fock (HF) method (B3LYP/6-31(d)), again using the programme Gaussian09 [13]. HF is another computational method based on quantum mechanical calculations, finding the wave function and hence energy of molecules. The HF method is more computationally expensive than DFT, but provides more accurate models for the ESP.

The generated ESPs were visualised at 99.8% of total electron density of the species using the programme PyMOL [15], with areas with a positive electrostatic potential coloured blue and those with a negative electrostatic potential coloured red, through white for neutral areas. The scale used for representation of ESPs was chosen to be appropriate to highlight differences between C–H protons and aromatic π-systems. This means that the most electropositive and electronegative regions of the molecules, which were the hydroxyl protons and oxygens, respectively, fall

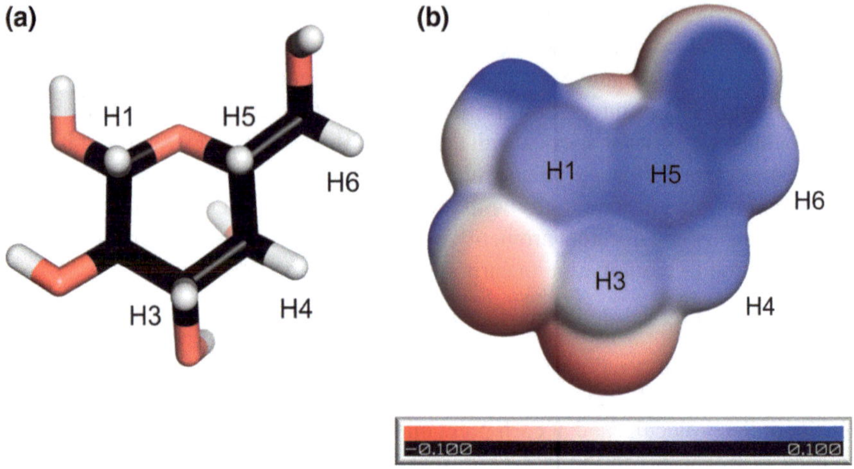

Fig. 3.5 Sample monosaccharide ESP with visible C–H protons numbered systematically. **a** Minimised conformation of β-D-Gal viewed from α-face. **b** Calculated ESP for minimised conformation of β-D-Gal, viewed from α-face. Electropositive regions coloured blue ($\geq +260\,\text{kJmol}^{-1}$), neutral white, and electronegative regions coloured red ($\leq -260\,\text{kJmol}^{-1}$). Scale shown in Hartrees (0.1 Hartrees $\approx 260\,\text{kJmol}^{-1}$); the same scale is used in all further ESP representations

beyond the range used. These calculations showed that the C–H protons do have a positive electrostatic potential, and are far from neutral or 'apolar' (Fig. 3.5b).

The Electronic Properties of Aromatic Species

The aromatic moieties of the amino acids were modelled by the same procedure, by generation of ESPs from minimised structures. Only the aromatic rings themselves were modelled; the ESPs of these differed very little when compared to the rings from calculations of complete modelled amino acids. When visualised in the same way as the monosaccharide ESPs, these showed the partial negative charge of the aromatic π-system (Fig. 3.6a).

Hydrogen bonding was an important consideration when modelling the aromatic rings—participation in hydrogen bonds will affect the π-system electronics, and these effects have been implicated in the relative preferences for involvement of amino acids in cation–π interactions, for example [16]. In the non-hydrogen-bonded ground state, the π-system of Tyr had a very similar electronegativity to that of Phe. However, in proteins Tyr acts as a hydrogen-bond donor through its hydroxyl in about 97% of cases [4]. This increases the electron density of the species, as the lone-pair electrons from the hydrogen-bond acceptor donate into the ring (Fig. 3.6b). In 40% of cases, the Tyr hydroxyl also acts as a hydrogen bond acceptor [4]. Although this leads to a reduction in the electronegativity of the π-system as the oxygen lone pair electrons donate into the bond, the overall effect is still an increased electon-density compared to the ground state (Fig. 3.6c).

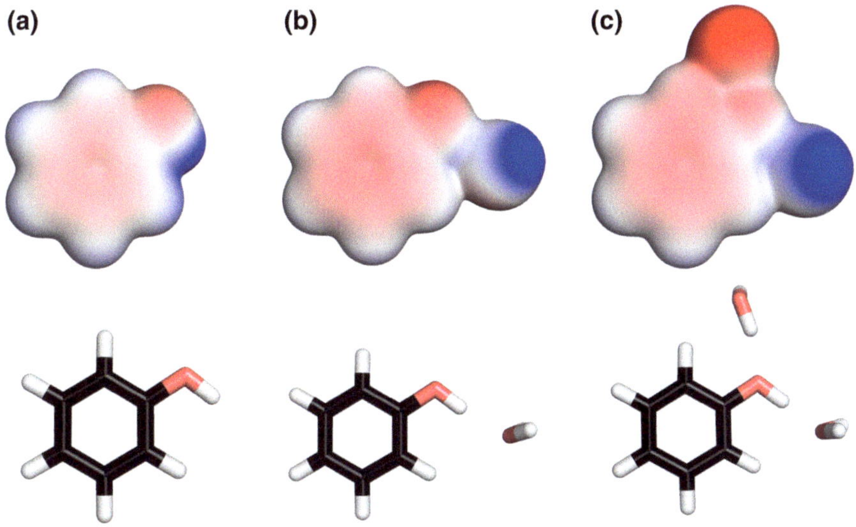

Fig. 3.6 Calculated ESPs for phenol molecules with optimised structures underneath: **a** Ground state; **b** As hydrogen-bond donor to water molecule; **c** As both hydrogen-bond donor and hydrogen-bond acceptor to water molecules. Electropositive regions coloured blue ($\geq +260\,\text{kJmol}^{-1}$), neutral white, and electronegative regions coloured red ($\leq -260\,\text{kJmol}^{-1}$)

The Trp ring N–H also acts as a hydrogen-bond donor in around 89% of cases [4], also increasing electron density in the π-system compared to the ground state. The ESPs that I used for consideration of Tyr and Trp were thus the most common forms, with each as single hydrogen-bond donors. The case for His is more complicated, as it can take multiple bonding forms, both as a neutral and protonated species; the ground state was used for comparison.

Participation in hydrogen-bonding through the hydroxyls and ring oxygen will also affect the electronics of the carbohydrates. However, this will be less pronounced for the C–H protons than for the aromatic rings given the larger separation to the bonds and the absence of direct mesomeric effects of the atoms concerned. I used the ground state ESPs for the carbohydrates, as the effect of hydrogen-bonding should be similar for all of the investigated carbohydrate species and so differences should be maintained. The calculated models of the electronics of the carbohydrates and aromatic rings enabled me to look for correlations with involvement in CH–π interactions.

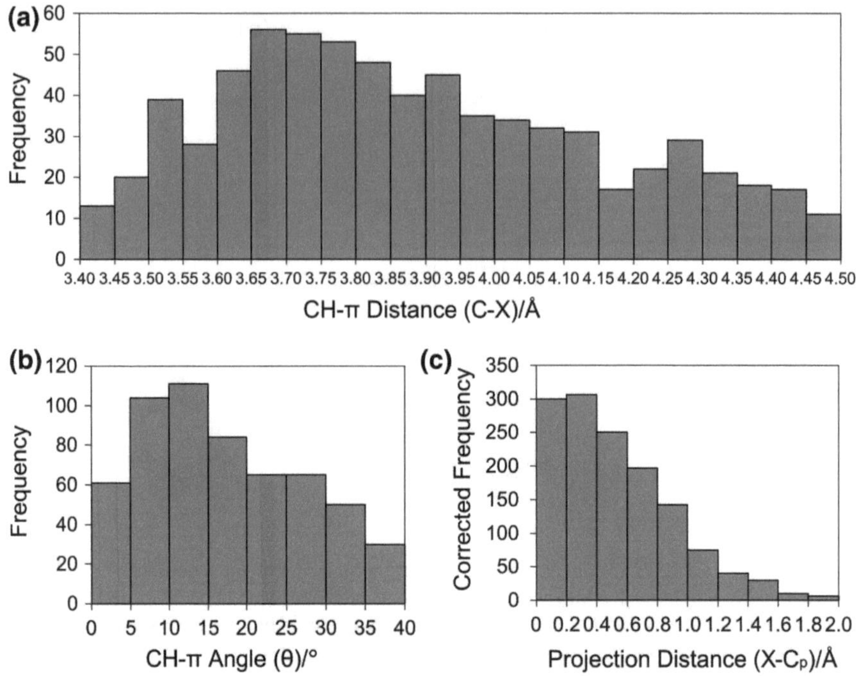

Fig. 3.7 Statistics for all CH–π interactions in data set involving investigated monosaccharides, identified using parameters in Fig. 2.7a. Adapted from Hudson et al. [20]. **a** CH–π distance. **b** CH–π angle. **c** C-projection distance, with cone correction [19]

3.4 Correlation Between Electrostatic Surface Potentials and CH–π Interactions

CH–π interactions were parametrised for the monosaccharide residues in the data set using the operational parametric definition (Fig. 2.7a). This identified a total of 721 aromatic rings participating in CH–π interactions (Table 3.1). The distribution of distances from the C–H carbon to the centre of the ring matched that seen in the previous study of such interactions within protein structures [17] (Fig. 3.7a), with an average of 3.88 Å. This value is close to the average distance between the C–H proton and plane of the ring found by a survey of the crystal structure database (2.75 Å) [18] plus the standard C–H bond length 1.09 Å.

The distribution of CH–π angles showed a minor preference for smaller angles, from 5° to 25° (Fig. 3.7b). For the C-projection distances, the frequencies had to be corrected for the areas represented by different bins: a larger projection distance represents a larger volume within which the carbon could be found. Therefore, the frequencies needed to be corrected for these different possible volumes [19]. Once corrected, shorter projection distances were preferred, with few CH–π interactions detected at the longer distances (Fig. 3.7c). This was likely partly due to limitations on the maximum distance possible with certain combinations of the other two parameters.

3.4.1 Participation of Amino Acids in CH–π Interactions

The aromatic amino acids did not participate equally in CH–π interactions with the investigated monosaccharide residues (Fig. 3.8a). The majority of interactions involved a Trp residue, and frequently each ring of the indole moiety interacted with a separate C–H bond (Fig. 2.7a). Tyr was the next most frequent CH–π acceptor, in about a quarter of cases, with Phe less commonly involved, and His the acceptor in very few cases. The trend for Trp, Tyr, and Phe reflects their relative propensities proximal to carbohydrates (Fig. 3.2), suggesting that these interactions may contribute to their over-representation. His has a high propensity despite being rarely involved in CH–π interactions, but it is able to interact in other ways—in the ground state it can act as both a hydrogen-bond donor and acceptor, and when protonated can participate in electrostatic interactions as well as be a hydrogen-bond donor.

The order of involvement as CH–π acceptors also followed the relative sizes of the amino acids [5]; Trp presents the largest aromatic ring and His the smallest. However, even when the frequencies as CH–π acceptors were corrected for the accessible surface areas of the amino-acid side chains the trend was maintained, with Trp accounting for over half of the residues involved in CH–π interactions (Fig. 3.8b).

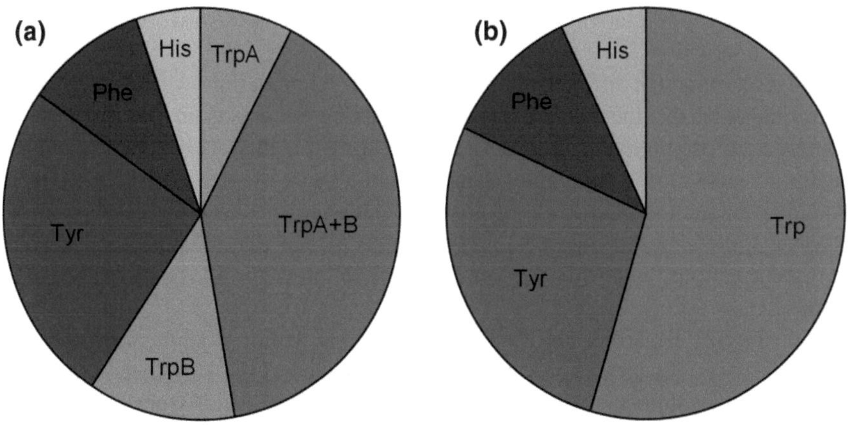

Fig. 3.8 Aromatic amino acids acting as acceptors in CH–π interactions with investigated monosaccharide residues in the data set. Reproduced from Ref. [20]. **a** Raw-count distribution. For Trp, interactions were identified with the five- or six-membered ring as TrpA and TrpB, respectively, or both, as TrpA+B. **b** Distribution normalised for accessible surface area of amino acid side chains [5]

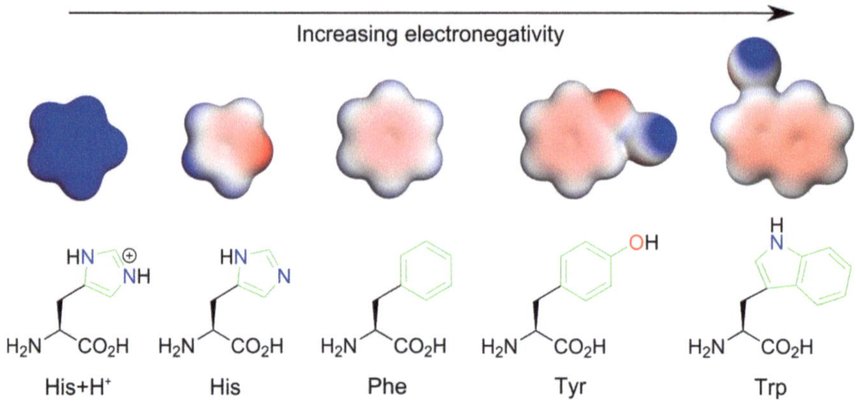

Fig. 3.9 Structures of aromatic amino acids with calculated ESPs for the corresponding aromatic rings (highlighted in green) shown above. Electropositive regions coloured blue ($\geq +260\,\mathrm{kJmol^{-1}}$), neutral white, and electronegative regions coloured red ($\leq -260\,\mathrm{kJmol^{-1}}$)

The involvement as CH–π acceptors did follow the relative electronic properties of the aromatic rings (Fig. 3.9). The indole ring of Trp is the most electron-rich aromatic system, and was the most common CH–π acceptor. The difference between Tyr and Phe is also instructive. Both amino acids have π-systems of almost equal size, and yet the previously discussed participation of the Tyr hydroxyl in hydrogen bonds means that its system is more electronegative than that of Phe. This explains why Tyr occurred much more often as a CH–π acceptor. In the ground state, His has the least electron-rich π-system, and it frequently acts as a hydrogen-bond acceptor [4] or can be protonated, which will reduce the electron-density of the π-system further. Corresponding to a relatively electron-poor π-system, His was involved least often in CH–π interactions.

The involvement of different amino acids in CH–π interactions with carbohydrates clearly indicates that the electronics of the aromatic system play an important role in carbohydrate–aromatic interactions. It is interesting that the relative participation of the amino acids closely resembles that found in a study of cation–π interactions in proteins [16]. As the interaction of positively charged species with electronegative π-systems, the cation–π interaction is analogous to the CH–π interaction. Indeed, for organic cations the positive charge often resides on the protons of polarised C–H bonds [21].

The study by Brandl et al. of CH–π interactions within proteins also finds that Trp is the most frequent CH–π acceptor, and His the least [17]. However, in that case Phe and Tyr are almost equally likely to participate. This differs from my findings, where for non-covalent carbohydrate–aromatic interactions Tyr is clearly a favoured acceptor over Phe. This may reflect differences for intramolecular interactions within proteins and intermolecular interactions with carbohydrate ligands: Within proteins, the aromatic rings must contact another group, and CH–π interaction may still represent the best interaction of a Phe side chain even if it is not as energetically favourable as the equivalent interaction with Tyr. However, for ligand binding this difference in energy for interaction with the two side chains could be vital for the binding to occur at all, hence why Tyr is favoured.

3.4.2 Distribution of CH–π Acceptors Around Carbohydrates

With the correlation between the electronics of the aromatic ring and participation in CH–π interactions, I then investigated whether the carbohydrate electronics also influenced these interactions. The different monosaccharide residues were involved in CH–π interactions to different extents (Table 3.1), despite all presenting similar numbers of C–H bonds (mostly seven, with Fuc presenting one more and Xyl one fewer). There were on average 0.66 CH–π interactions per example. α-D-Man participated the least often with an average of 0.22 interactions, ranging to 1 interaction per example for β-D-Gal, or 1.64 for α-L-Fuc, although this was from only 11 examples. In all cases, the β-anomers were involved in CH–π interactions more frequently than the α-anomers, and the *N*-acetylamino sugars had fewer interactions than their parents (Table 3.1).

CH–π Interactions of β-D-Gal

β-D-Gal is illustrative of the effects of carbohydrate electronics on CH–π interactions (Fig. 3.10), as the well-represented residue most frequently involved in CH–π interactions. The calculated ESP for β-D-Gal shows a distinct electropositive patch on the α-face corresponding to C(1)–H, C(3)–H, and C(5)–H (Fig. 3.10b). This extends around the side where C(4)–H and one C(6)–H (in this minimised, and most common, conformation) are situated. The electrostatic potential for the C(2)–H and the remaining C(6)–H on the β-face are noticeably less positive.

The positions of the amino-acid aromatic rings identified as participating in CH–π interactions with β-D-Gal in the data set were visualised (Fig. 3.10c). The rings were concentrated on the α-face and around the side corresponding to the electropositive patch seen in the ESP. Very few aromatic rings were identified interacting with the less electropositive C–H protons on the β-face.

The propensity for each carbohydrate C–H bond to participate in CH–π interactions was quantified by counting the number of rings identified interacting with each bond (Sect. 2.3) and normalising for the number of examples, and number of C–H bonds in the case of C6 (Fig. 3.10d). This confirmed that the most electropositive C(5)–H proton was most often involved in interactions, with almost as many involving C(4)–H. These were also the two C–H bonds for which the interacting side chains were almost exclusively aromatic (Fig. 3.4). C(1)–H, C(3)–H, and C(6)–H also participated in CH–π interactions, although less often at around one in ten cases. The least electropositive C(2)–H was confirmed as forming CH–π interactions rarely, if ever.

The relative involvements of the β-D-Gal C–H bonds in CH–π interactions were consistent with the contribution of the carbohydrate electronics to carbohydrate–

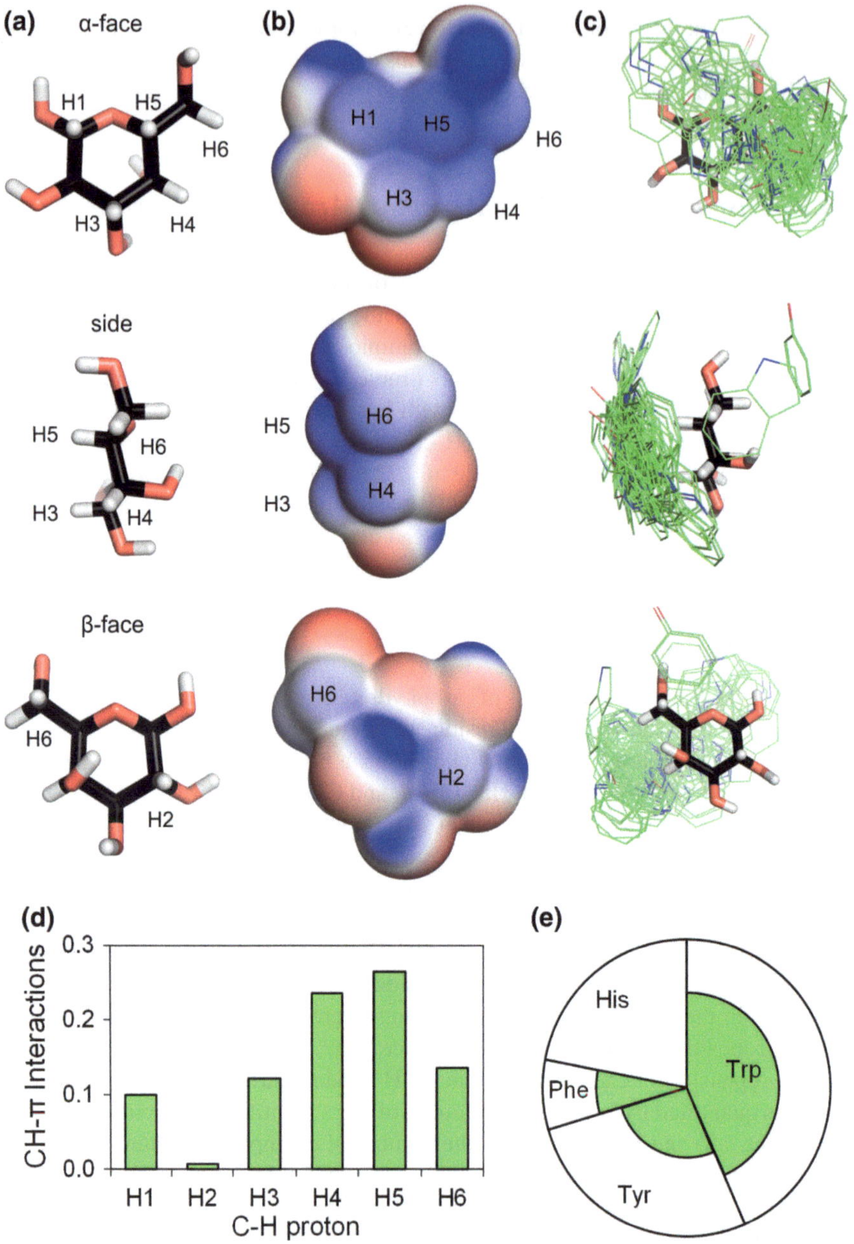

Fig. 3.10 CH–π interactions involving β-D-Gal in data set. Adapted in part from Hudson et al. [20]. **a** Three views of optimised structure. **b** ESP calculated for optimised structure. **c** All monosaccharide residues in data set overlaid, with identified amino acids CH–π acceptor aromatic rings shown in green. **d** Average involvement of C–H protons in CH–π interactions. **e** White pie chart: proximal aromatic amino acids; Green wedges: those involved in CH–π interactions

aromatic interactions. However, as well as being the most neutral, the C(2)–H is also in a hindered position next to the axial C(4)–OH, which may impede interactions with bulky aromatic moieties. The analysis of further monosaccharides confirmed the role of electronics.

CH–π Interactions of Further Monosaccharides

β-D-Glc was well-represented in the data set, and was involved in CH–π interactions less frequently than β-D-Gal at an average of 0.8 interactions per example (Table 3.1). The all-equatorial configuration of β-D-Glc means that the α-face is not sterically hindered (Fig. 3.11a), similar to that of β-D-Gal. However, the different configuration leads to C–H protons on the α-face with reduced electropositivity for β-D-Glc (Fig. 3.11b). The aromatic rings involved in CH–π interactions were mostly found over the centre on the α-face, reflecting the shape of the electropositive patch formed by these C–H bonds (Fig. 3.11c). The C(5)–H and α-face C(6)–H protons, for which the electrostatic potential differs the most for β-D-Glc compared to β-D-Gal, were correspondingly less frequently involved in CH–π interactions, although interactions were more frequent for C(1)–H (Fig. 3.11d).

The C–H protons on the β-face of β-D-Glc were involved in CH–π interactions more frequently than was C(2)–H of β-D-Gal (Fig. 3.11c); β-D-Glc has the more electropositive β-face. However, the interactions were reduced for C(4)–H in β-D-Glc compared to those observed for the more electropositive C(4)–H in β-D-Gal.

Compared to β-D-Glc, α-D-Glc has fewer interactions for C(3)–H and C(5)–H on the α-face (Fig. 3.12c, d) because of reduced electropositivity of this face due to the axial C(1)-OH (Fig. 3.12b). This is the only axial C–OH of α-D-Glc and is on the α-face, the opposite case to that of β-D-Gal, and the C–H bonds that participated most often in CH–π interactions were on the β-face, also opposite to β-D-Gal. On the β-face the C–H protons are equally unhindered as their equivalents for β-D-Glc, but for α-D-Glc they were more frequently involved in interactions corresponding to increased electropositivity.

The comparison of the two anomers of Gal is analogous to that for Glc. The additional axial C–OH on the α-face reduces the electropositivity for α-D-Gal relative to β-D-Gal (Fig. 3.13b), with a reduction in the observed participation in CH–π interactions of C(3)–H and C(5)–H (Fig. 3.13c, d). The β-face C–H protons for α-D-Gal were also rarely involved in interactions due to the C(4)–OH on that face.

Fuc provided a useful comparison to Gal, as it is equivalent to 6-deoxy-Gal. It is most common in nature as the L-form, the mirror image of the D-form prevalent for Gal, but as D- and L-forms are enantiomers the electronic properties are not affected. Fuc was most frequent in the data set as the α-anomer (Table 3.1), which can be compared to α-D-Gal; there were insufficient examples of β-L-Fuc to draw conclusions. As expected, the aromatic rings involved in CH–π interactions for α-L-Fuc formed a shape approximately a mirror image to those for α-D-Gal (Fig. 3.14c), but average participation in CH–π interactions was substantially reduced for the C–H bonds, particularly C(5)–H and C(6)–H (Fig. 3.14d). The lack of a C(6)–OH

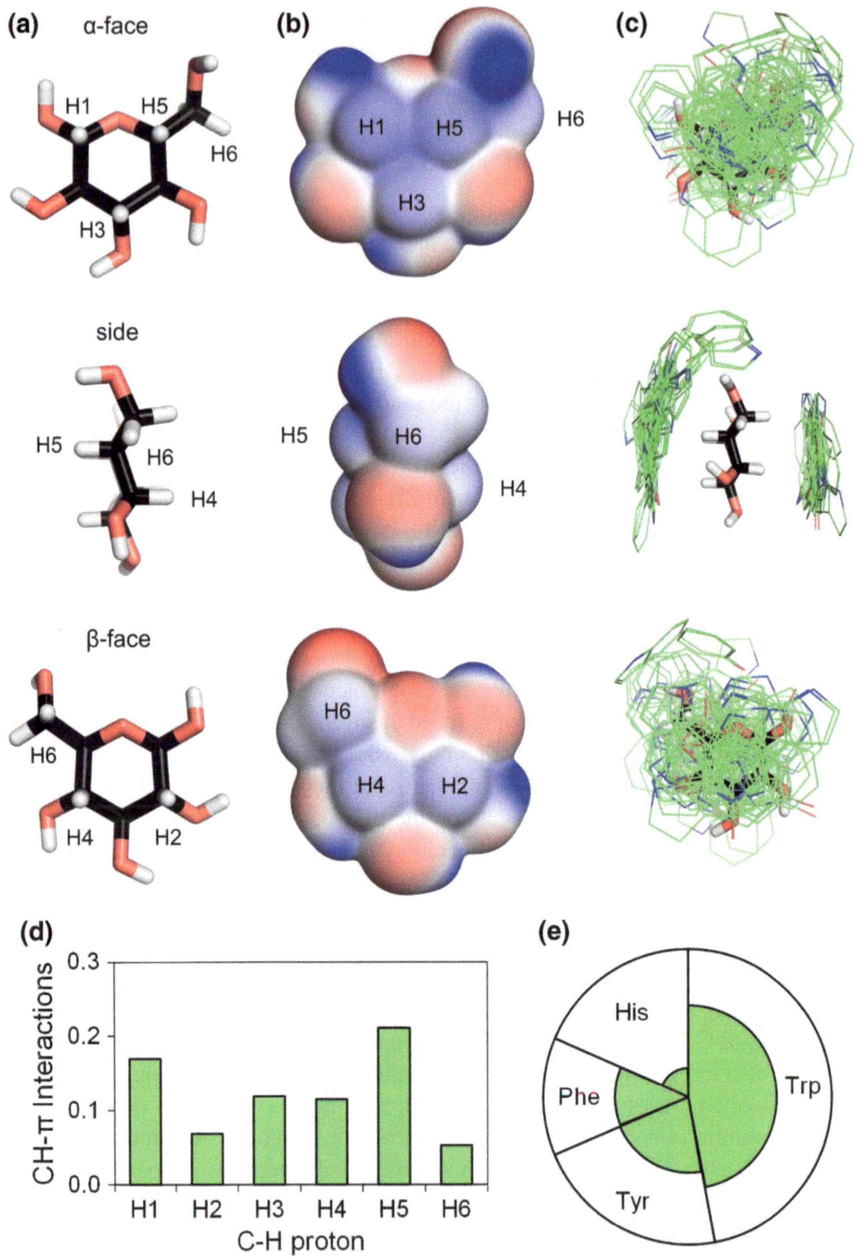

Fig. 3.11 CH–π interactions involving β-D-Glc in data set. Adapted in part from Hudson et al. [20]. **a** Three views of optimised structure. **b** ESP calculated for optimised structure. **c** All monosaccharide residues in data set overlaid, with identified amino acids CH–π acceptor aromatic rings shown in green. **d** Average involvement of C–H protons in CH–π interactions. **e** White pie chart: proximal aromatic amino acids; Green wedges: those involved in CH–π interactions

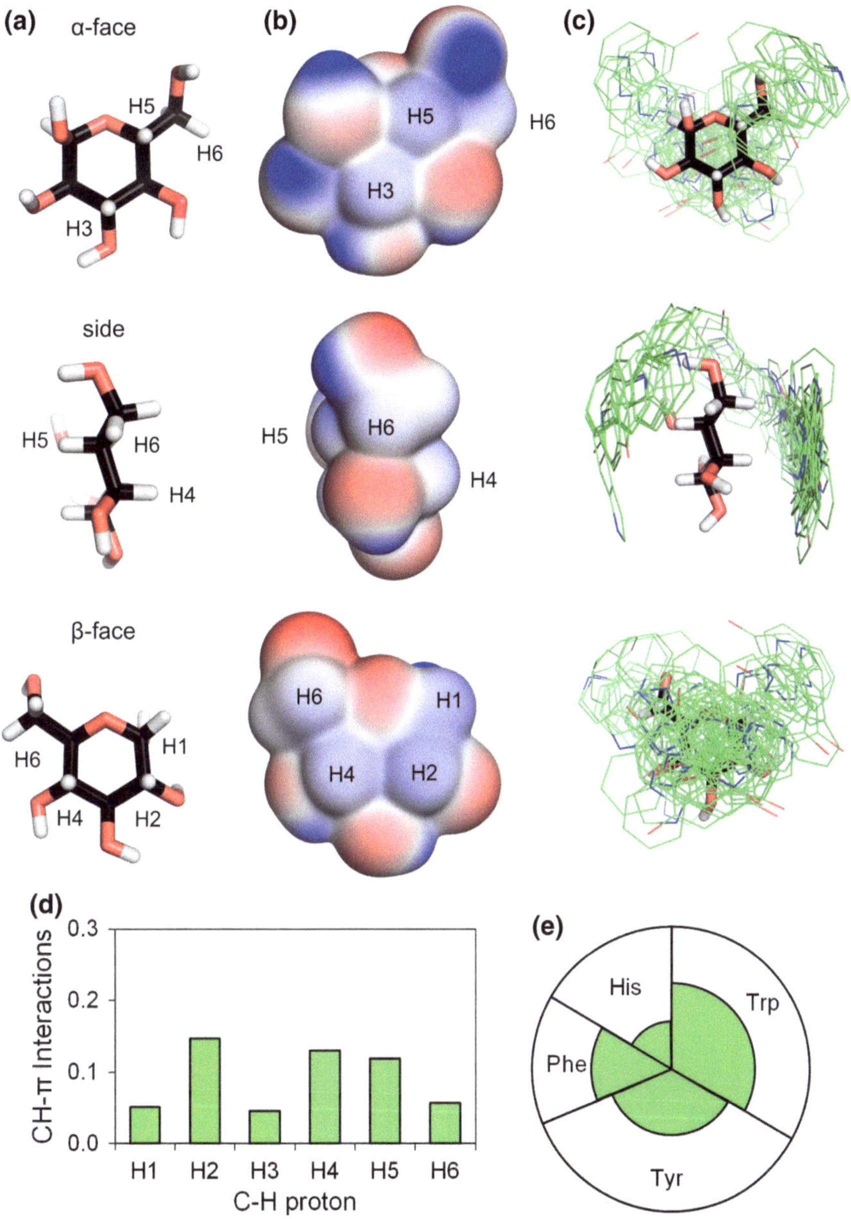

Fig. 3.12 CH–π interactions involving α-D-Glc in data set. Adapted in part from Hudson et al. [20]. **a** Three views of optimised structure. **b** ESP calculated for optimised structure. **c** All monosaccharide residues in data set overlaid, with identified amino acids CH–π acceptor aromatic rings shown in green. **d** Average involvement of C–H protons in CH–π interactions. **e** White pie chart: proximal aromatic amino acids; Green wedges: those involved in CH–π interactions

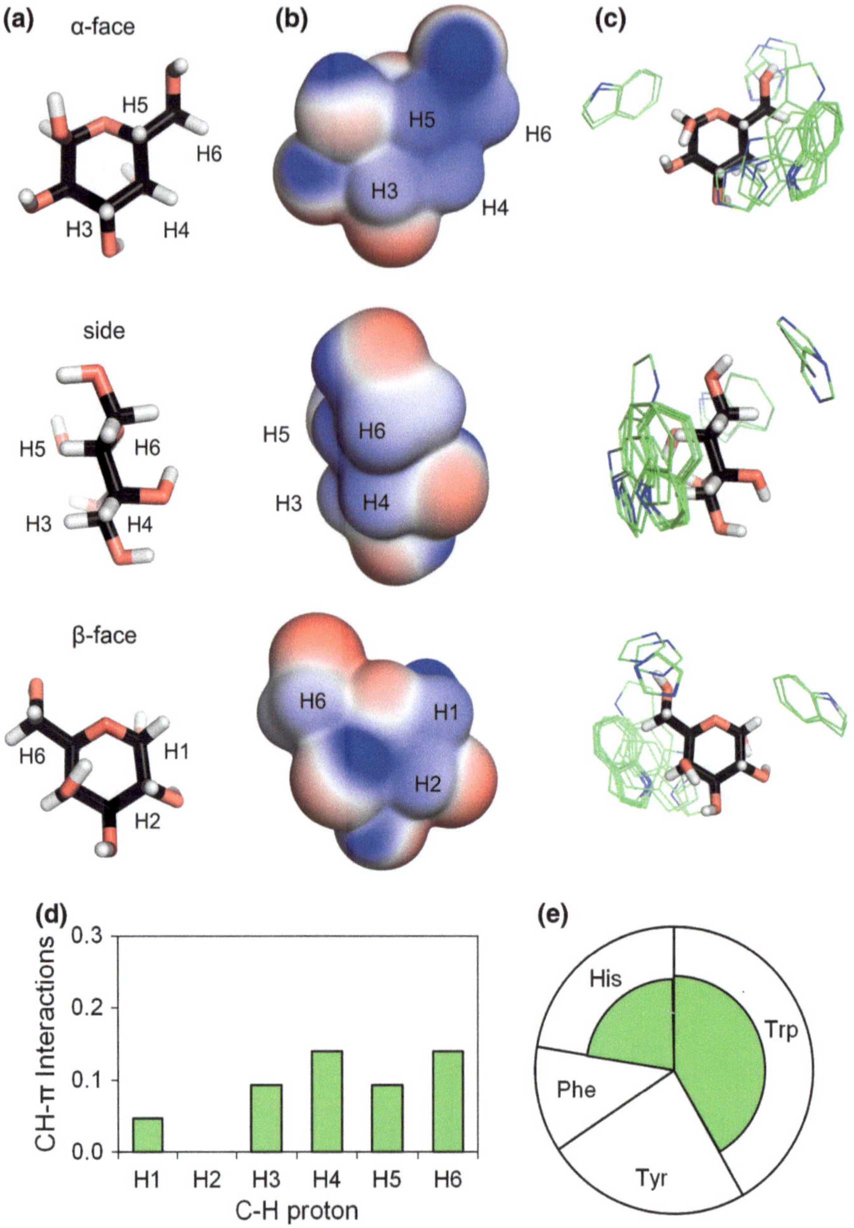

Fig. 3.13 CH–π interactions involving α-D-Gal in data set. **a** Three views of optimised structure. **b** ESP calculated for optimised structure. **c** All monosaccharide residues in data set overlaid, with identified amino acids CH–π acceptor aromatic rings shown in green. **d** Average involvement of C–H protons in CH–π interactions. **e** White pie chart: proximal aromatic amino acids; Green wedges: those involved in CH–π interactions

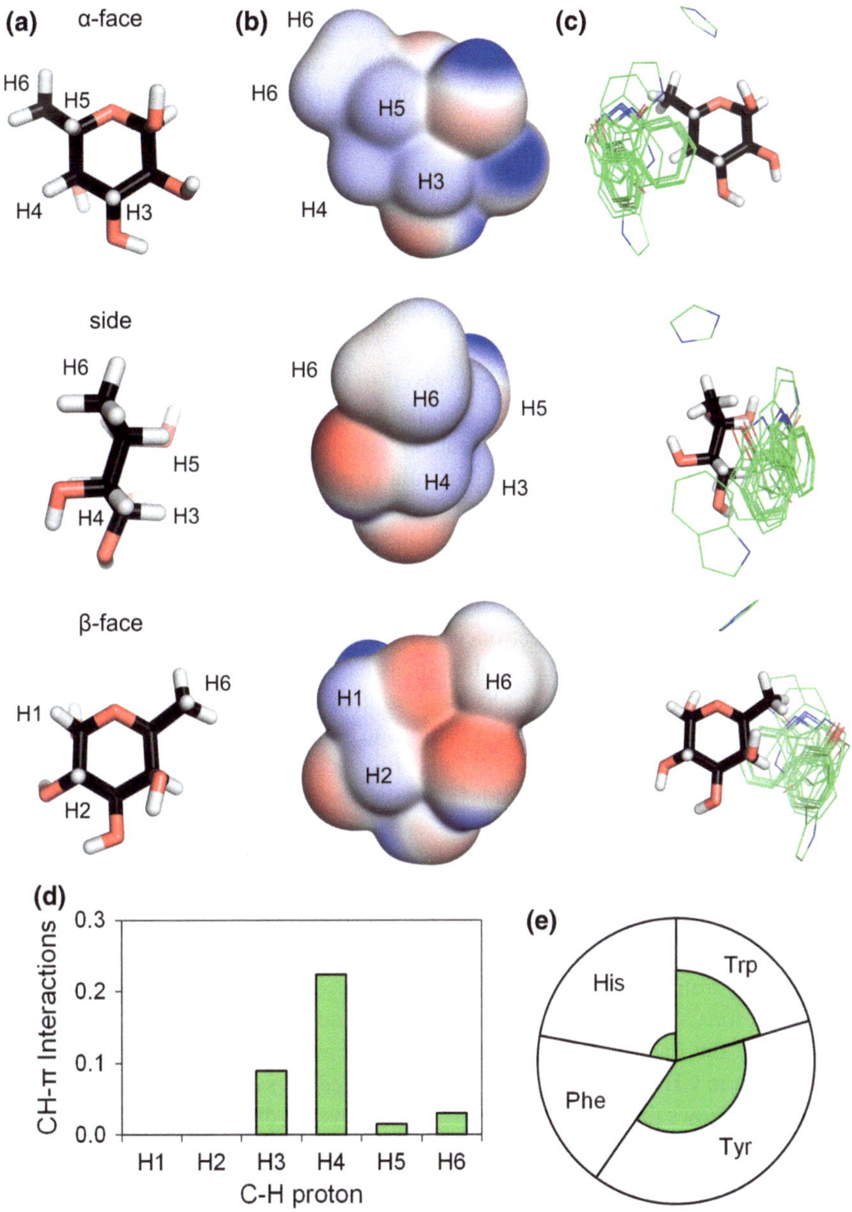

Fig. 3.14 CH–π interactions involving α-L-Fuc in data set. **a** Three views of optimised structure. **b** ESP calculated for optimised structure. **c** All monosaccharide residues in data set overlaid, with identified amino acids CH–π acceptor aromatic rings shown in green. **d** Average involvement of C–H protons in CH–π interactions. **e** White pie chart: proximal aromatic amino acids; Green wedges: those involved in CH–π interactions

for α-L-Fuc results in less electropositive C–H protons, particularly at C6, compared to α-D-Gal (Fig. 3.14b). This shows the importance of electronics in carbohydrate–aromatic interactions over a general hydrophobic effect—due to its methyl group Fuc is more hydrophobic than Gal, and yet is involved in CH–π interactions less frequently.

The relative involvement in CH–π interactions of the C–H bonds of the other monosaccharide residues that were investigated supported the findings illustrated with the above examples (Sect. 7.2.3). Man is similar to Gal in that it has an axial C–OH on the β-face, which is at the C2 rather than C4 position. This means that β-D-Man has an analogous electropositive patch to that of β-D-Gal, but for Man it is formed of C(1)–H, C(2)–H, and C(3)–H (Fig. 7.7). The examples of β-D-Man in the data set were found frequently to be involved in CH–π interactions on the α-face, with most interactions involving C(1)–H, but also C(5)–H. Unlike for Gal, for the α-anomer of Man this electropositive patch is disrupted by the axial C(1)–OH, and α-D-Man has no C–H protons with a particularly positive electrostatic potential and was not involved in many CH–π interactions (Fig. 7.6).

For the *N*-acetyl amino sugars, the main difference observed from the parent monosaccharides was a reduction in CH–π interactions at the ring C–H bonds near to the *N*-acetyl amino groups on C2 (Sect. 7.2.3). A reduction in the electropositivity of the C–H protons near to the electronegative carbonyl oxygen is evident in the ESPs. β-D-GlcNAc participated in CH–π interactions particularly rarely compared to β-D-Glc (Fig. 7.5), which may explain why electrostatics were found to be unimportant in a study of carbohydrate–aromatic interactions using covalently linked β-D-GlcNAc [22]. Interactions involving the C–H bonds of the acetyl group were often seen, and these are the most positive C–H bonds in the species, but these fell outside the scope of this study.

Together, the data for all of the monosaccharides show a clear correlation between the electrostatic potentials of carbohydrate C–H protons and their involvement in CH–π interactions within protein X-ray crystal structures. Particularly electropositive C–H protons are most-frequently engaged in CH–π interactions, as are the most-electron-rich aromatic amino acids. Steric hindrance from C–OH groups may also play a role for C–H bonds that do not engage frequently in such interactions. For cases where the only differences between C–H environments are the electronics participation in CH–π interactions is directly related to positivity. The data for Fuc in particular shows that electronics are dominant over hydrophobic effects.

These trends are indicative of an important contribution of electrostatics to carbohydrate–aromatic interactions. I carried out experiments to probe such non-covalent interactions in solution to determine the extent of this electrostatic contribution.

3.5 Studies of Carbohydrate–Aromatic Interactions in the Solution Phase

Acknowledgement: The work presented in this section was carried out with Roger Diehl in the lab of Prof. Laura Kiessling (University of Wisconsin-Madison, Madison, WI, USA)

I investigated the influence of electrostatics on carbohydrate–aromatic interactions by ^{1}H NMR spectroscopy of glycosides in D_2O. Methyl glycosides were used as models for the carbohydrates, as simple monosaccharide residues that would not anomerise. I chose indole as the aromatic species, as a substitute for Trp given the prevalence of this amino acid as an acceptor of CH–π interactions in the crystallographic data set. Similar studies have been published [23], particularly two using methyl glycosides in solution mixed with benzene, phenol, Phe, and Trp, [24, 25] upon which this part of the study was based.

The chemical shifts of C–H protons directly report participation in CH–π interactions, with a positive change in δ (Δδ) in the presence an aromatic species due to the formation of a CH–π interaction and the magnitude indicative of the interaction strength (Sect. 1.2.4) [24]. Therefore, I measured the ^{1}H NMR spectra of the methyl glycosides in the absence and presence of indole, to obtain the Δδ values for the C–H protons to report on carbohydrate–aromatic interactions.

3.5.1 *Influence of C–H Electronics on CH–π Interactions*

β-D-Gal and β-D-Glc were selected as they were both prone to engage in CH–π interactions in the crystallographic data set, with C–H groups with different properties and propensities to engage in these interactions. Methyl-β-D-Man was also synthesised (no synthesis of the deprotected form has been published previously), as a monosaccharide for which I could predict the modes of interaction, but for which there were insufficient examples in this data set for confirmation.

The Δδ values differed for the various glycoside C–H bonds (Fig. 3.15). The largest Δδ values were around -5 ppb (Fig. 3.15d), representing changes of 2–3 Hz on a 500 MHz spectrometer, while for some protons no changes were observed (Fig. 3.15e). The small changes in chemical shift show transient and weak interactions, but they were measurable and consistent. The carbohydrate C–H protons engaged in CH–π interactions with the indole, but to different extents within and between monosaccharides, as I had found in the crystallographic data set.

The Δδ values depended on indole concentration, and all were linearly related up to the solubility limit of indole in D_2O (Fig. 3.16). For β-D-Gal, the relative Δδs upon addition of indole reflected the positivity of the calculated electrostatic potentials at the different C–H protons, as well as their relative involvement in CH–π interactions in the crystallographic database (Figs. 3.10b, d and 3.16a). The most electropositive H5 had the largest change in chemical shift, while the less-positive H2

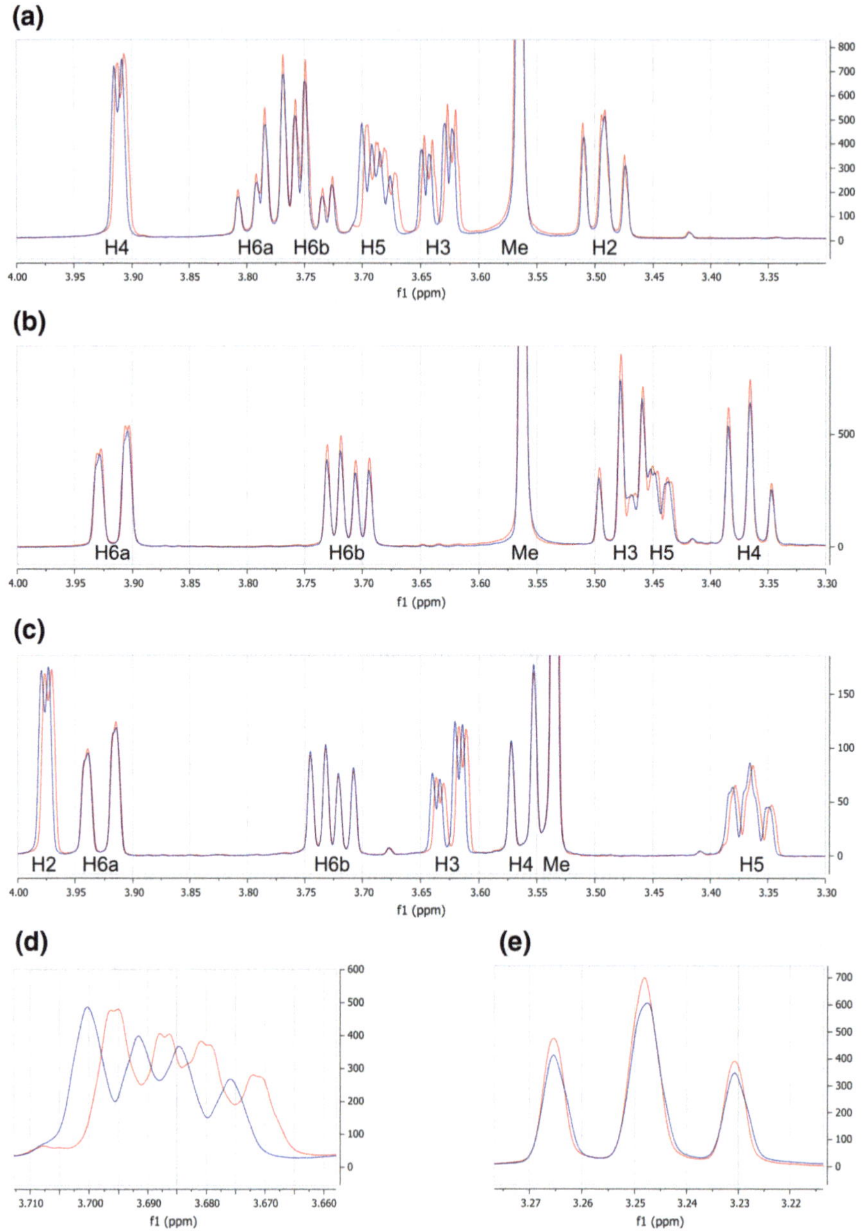

Fig. 3.15 Representative NMR spectra for methyl glycosides. C–H proton region of ^{1}H NMR spectra for 10 mM methyl glycoside solutions in D_2O with (red) and without (blue) 7.5 mM indole, with proton peaks labelled. Adapted from Hudson et al. [20]. **a** β-Methyl-D-Gal. **b** β-Methyl-D-Glc. **c** β-Methyl-D-Man. Expanded views shown for **d** β-Methyl-D-Gal, showing -5 ppb δΔ, and **e** H4 of β-Methyl-D-Glc, showing 0 ppb δΔ

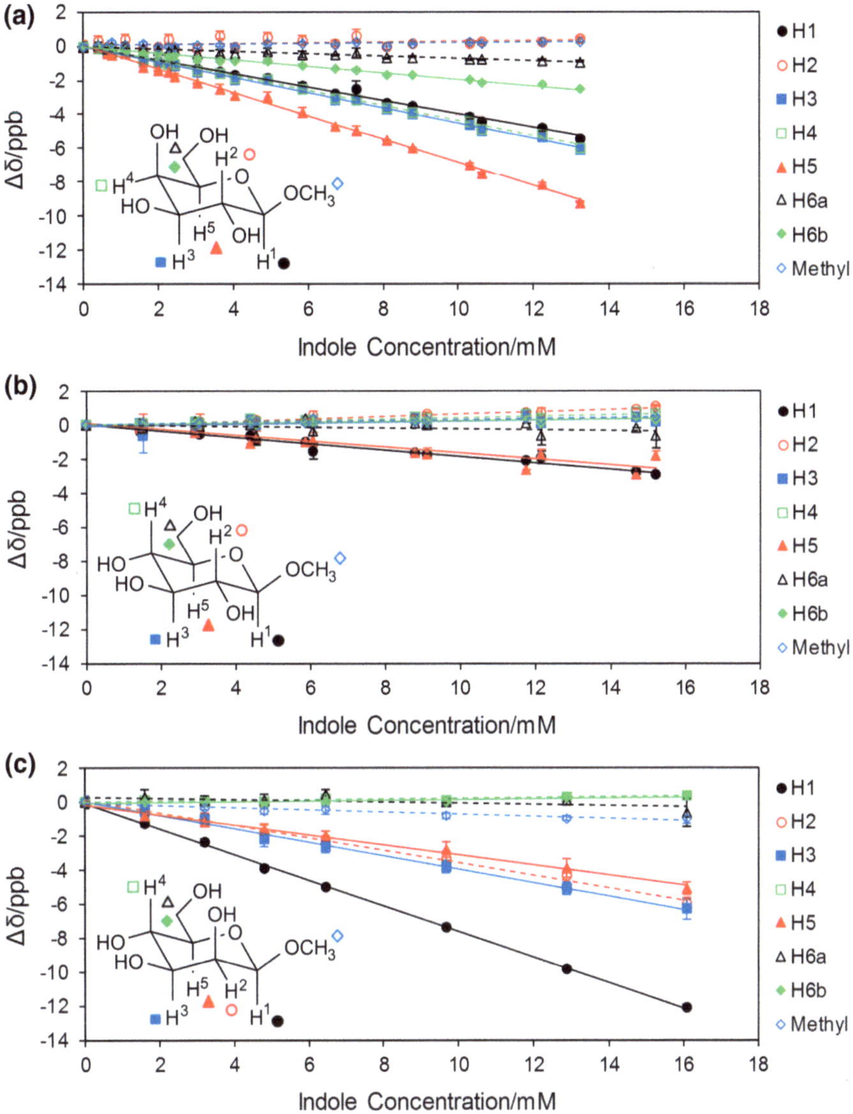

Fig. 3.16 Changes in δ (Δδ) of C–H protons of methyl glycosides by ^{1}H NMR spectroscopy at different concentrations of indole, relative to indole-free solutions. Adapted from Hudson et al. [20]. **a** β-D-Gal. **b** β-D-Glc. **c** β-D-Man. Linear fits shown as lines of same colour as markers, solid for solid markers, dashed for hollow markers. Error bars are standard deviation

did not change, indicative of it not participating in CH–π interactions, as I found in the protein structures. H1, H3, and H4 saw smaller changes, interacting less strongly with the aromatic species. The methyl group served as a useful control—the C–H protons were not expected to participate in interactions with the indole and the chemical shift was not affected by indole concentration.

In comparison to β-D-Gal, β-D-Glc showed much smaller Δδ values upon addition of indole, consistent with it having less-electropositive C–H bonds and weaker CH–π interactions (Figs. 3.11b and 3.16b). The only C–H protons that showed a negative Δδ with increased indole concentration were the most elecropositive of β-D-Glc, H1 and H5. Some β-D-Glc C–H protons, such as H2, showed an unexpected positive Δδ with indole, which could be due to a side-on interaction as a result of hydrogen bonding between the indole and the glycoside.

β-D-Man C–H protons showed responses to indole of a similar magnitude to those of β-D-Gal (Fig. 3.16c), as predicted due to the analogous electropositive C–H patches resulting from the axial hydroxyl group. For β-D-Man, the biggest Δδ values were seen for H1, at the anomeric position. Smaller responses were seen for the other α-face C–H bonds, with no observable response for H4 on the β-face or the methyl protons.

The δ responses of these methyl glycosides show differences in the formation of CH–π interactions for C–H bonds in different electrostatic environments. This supports a role for electrostatics in non-covalent carbohydrate–aromatic interactions, reinforcing my findings from the survey of protein X-ray crystal structures. The results for β-D-Man show that the trends can be used to predict the interactions of carbohydrates not well represented in the bioinformatics analysis. The data also agree with relative changes in chemical shift of different protons that have previously been published for methyl glycosides interacting with benzene or phenol [24] (Table 3.2). When scaled to the same concentration of aromatic species, the magnitudes of Δδ for the three species also correlate with the preferred CH–π acceptors in the data set, with indole giving the largest changes.

3.5.2 Influence of Aromatic Electronics on CH–π Interactions

To further examine the role of the electronics of the aromatic species on CH–π interactions, ^{1}H NMR studies were carried out with the methyl glycosides and a range of 5-substituted indoles. The indole substituents were chosen to have different effects on the electronics of the aromatic ring, from the mesomerically electron-donating hydroxy group to inductively electon-withdrawing groups including cyano and nitro; these effects are quantified using Hammett σ_p parameters [26]. The effect of the substituent on the electronics of the aromatic species was quantified and visualised by the calculation of ESPs for the 5-substituted indoles (Fig. 3.17). They showed that the substituent altered the aromatic electronics, meaning that the indoles used

Table 3.2 Comparison of Δδ recorded for β-D-Gal at 7.5 mM indole with published values with benzene and indole (500 MHz spectrometer) [24]. Δδ is linearly dependent on phenol concentration beyond 100 mM [25]. Adapted from Hudson et al. [20]

C–H Proton Aromatic species concentration	**Δδ/Hz**		**Δδ (linear scaled)/ppb**		
	Benzene 10 mM	Phenol 100 mM	Benzene 7.5 mM	Phenol 7.5 mM	Indole 7.5 mM
H1	−0.5	−4.0	−0.8	−0.6	−2.8
H2	0.3	3.0	0.5	0.5	0.2
H3	−0.6	−3.7	−0.9	−0.6	−3.1
H4	−0.6	−3.6	−0.9	−0.5	−3.1
H5	−1.0	−7.3	−1.5	−1.1	−4.8
H6a	0.0	0.9	0.0	0.1	−0.5
H6b	−0.3	−0.8	−0.5	−0.1	−1.4

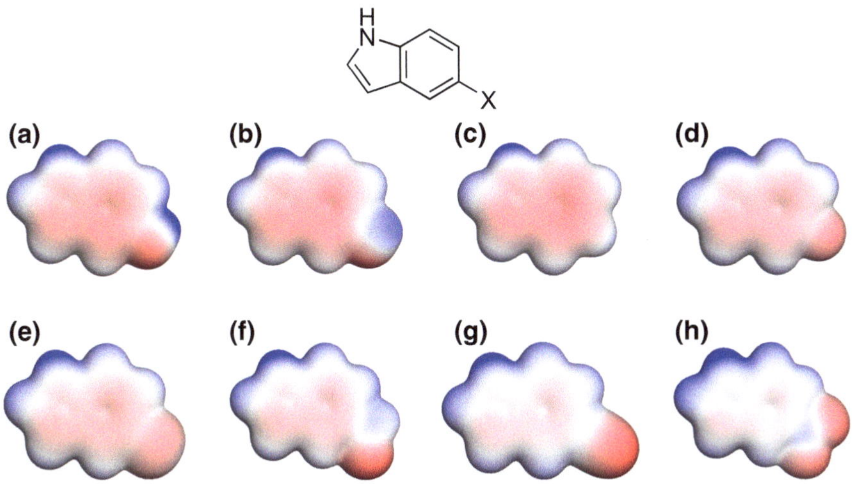

Fig. 3.17 Calculated ESPs for 5-substituted indoles used in Hammett-type analysis. **a** 5-Hydroxyindole, X = OH (−0.38). **b** 5-Aminoindole, X = NH_2 (−0.30). **c** Indole, X = H (0.00). **(d)** 5-Fluoroindole, X = F (0.15). **e** 5-Chloroindole, X = Cl (0.24). **f** 5-Formylindole, X = COH (0.42). **g** 5-Cyanoindole, X = CN (0.66). **(h)** 5-Nitroindole, X = NO_2 (0.81). Hammett σ_p value [26] of 5-substituent in parentheses. Calculated as non-hydrogen-bonded forms for simplicity

presented surfaces with electrostatic potentials ranging from fairly electronegative to almost neutral. Comparing the changes in C–H proton shifts of glycosides interacting with these indoles to those in indole-free solution was then performed by a Hammett-type analysis [27].

Δδ values were measured for each indole relative to indole-free solutions for the glycosides at 7.5 mM indole. In some cases limited solubility of the indole meant that this value was obtained by linear scaling, for example for 5-nitroindole. Plotting

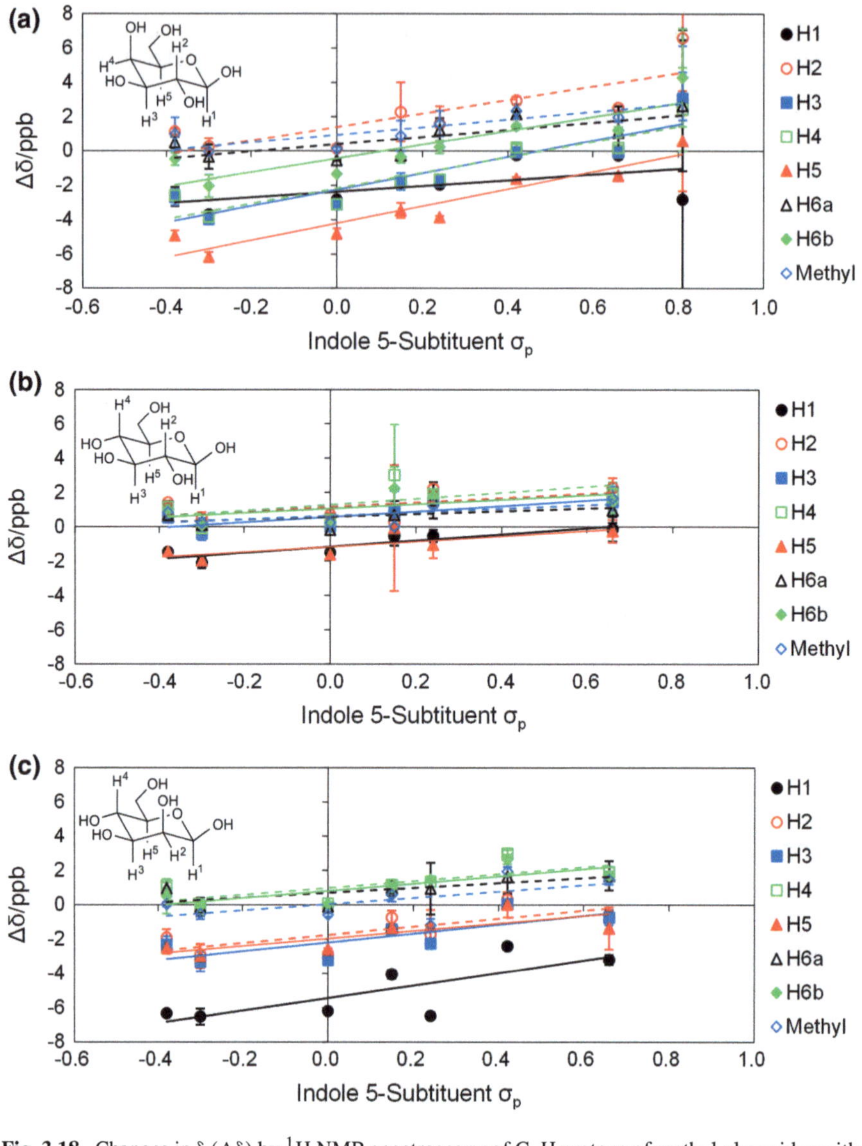

Fig. 3.18 Changes in δ (Δδ) by ^{1}H NMR spectroscopy of C–H protons of methyl glycosides with 5-substituted indoles relative to indole (all scaled to 7.5 mM indole concentration). Adapted from Hudson et al. [20]. **a** β-D-Gal. **b** β-D-Glc. **c** β-D-Man. Linear fits shown as lines of same colour as markers, solid for solid markers, dashed for hollow markers. Error bars are standard deviation

the Δδ values for the different glycoside C–H protons against the σ_p values of the 5-substituent showed the influence of electronics on CH–π interactions in a similar way to a Hammett plot (Fig. 3.18). The protons with Δδ values indicative of the formation of CH–π interactions responded to the aromatic electronics, with larger values seen with more electron-rich aromatics. For example, the H5 of β-D-Gal shows a large Δδ with the electron-rich indoles, with reductions in the value with less electron-rich species, suggesting weaker interactions (Fig. 3.18a). For 5-nitroindole, with an almost neutral electrostatic potential over the aromatic ring, no change is observed at all, showing that electronics are key to the carbohydrate–aromatic interaction.

Similar trends were observed for all of the carbohydrate C–H protons that had altered chemical shifts in solutions with indole. Those with no Δδ with indole, such as the methyl group of β-D-Gal, did not show any interaction even with the most electron-rich indoles (Fig. 3.18a). However, these protons did show increased positive Δδ values with the less-electron-rich indoles. This was unexpected, as this suggests a mechanism of altering chemical shift other than the formation of CH–π interactions. One possible explanation is that the positive Δδ, i.e., downfield change in shift, is due to edge-on effects from the indoles hydrogen bonding to the carbohydrate hydroxyls. These may be counteracted for the more electron-rich indoles through the formation to weak CH–π interactions that do not form with the less-electron-rich derivatives. It is also possible that the response is a non-specific effect of the aromatic species in the solution.

Comparing the gradients of the plots against σ_p for the C–H protons showed that a dependence on the magnitude of the Δδ (Fig. 3.19), i.e., that the protons with stronger interactions with the aromatic species showed a greater response to the aromatic electronics. This implies that the electronics do have an effect through the formation of CH–π interactions, although this may appear enhanced due to an additional global effect.

The results obtained from the NMR spectroscopy experiments confirm the role of electrostatics in carbohydrate–aromatic interactions. Even for the weak and transient

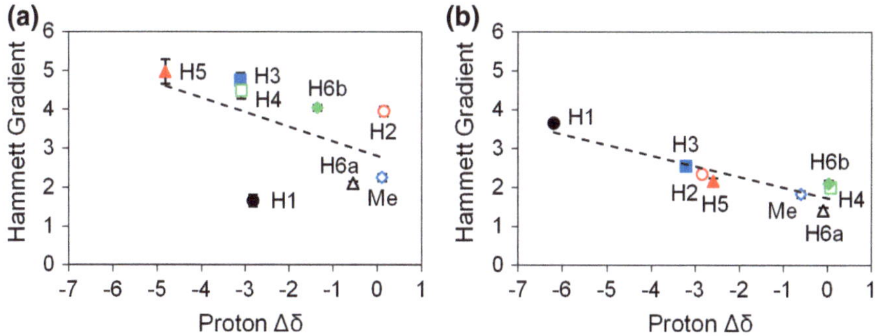

Fig. 3.19 Correlation between Δδ of glycoside C–H protons with 7.5 mM non-substituted indole and gradient of Hammett plot for that proton. **a** β-Methyl-D-Gal. **b** β-Methyl-D-Man. Linear fits shown as dashed lines. Error bars are standard deviation

interactions between free glycosides and aromatic rings in solution, stronger interactions are observed for C–H protons with a more positive electrostatic potential. The aromatic electronics are also important, and those with a neutral electrostatic potential do not interact with even the most positive C–H protons, showing that electrostatics are key to the interaction.

References

1. Fauchere JL, Pliska VE (1983) Eur. J. Med. Chem. 18:369–375
2. Bairoch A, Apweiler R, Wu CH, Barker WC, Boeckmann B, Ferro S, Gasteiger E, Huang H, Lopez R, Magrane M, Martin MJ, Natale DA, O'Donovan C, Redaschi N, Yeh L-SL (2005) Nucleic Acids Res. 33:D154–D159
3. UniProtKB/Swiss-Prot Protein Knowledgebase Release. http://web.expasy.org/docs/relnotes/relstat.html
4. McDonald IK, Thornton JM (1994) J. Mol. Biol. 238:777–793
5. Miller S, Janin J, Lesk AM, Chothia C (1987) J. Mol. Biol. 196:641–656
6. Lemieux RU (1996) Acc Chem Res 4842:373–380
7. Asensio JL, Ardá A, Cañada FJ, Jiménez-Barbero J (2013) Acc Chem Res 46:946–954
8. Perez S, Marchessault RH (1979) Biopolymers 18:2369–2374
9. Kirschner KN, Woods RJ (2001) Proc Natl Acad Sci USA 98:10541–10545
10. Nishio M (2011) Phys Chem Chem Phys 13:13873–13900
11. Weiner PK, Langridge R, Blaney JM, Schaefer R, Kollman PA (1982) Proc Natl Acad Sci USA 79:3754–3758
12. Cisneros GA, Karttunen M, Ren P, Sagui C (2014) Chem Rev 114:779–814
13. Frisch MJ et al. (2009) Gaussian 09, Revision D.01. 2009
14. Hohehnberg P, Kohn W (1964) Phys Rev 136:B864–B871
15. The PyMOL Molecular Graphics System, Version 1.7.0.3. (2014)
16. Mecozzi S, West AP, Dougherty DA (1996) Proc Natl Acad Sci USA 93:10566–10571
17. Brandl M, Weiss MS, Jabs A, Sühnel J, Hilgenfeld R (2001) J Mol Biol 307:357–377
18. Takahashi O, Kohno Y, Iwasak S, Saito K, Iwaoka M, Tomoda S, Umezawa Y, Tsuboyama S, Nishio M (2001) Bull Chem Soc Jpn 74:2421–2430
19. Ciunik Z, Desiraju GR (2001) Chem. Commun. (Cambridge, U.K.), 703–704
20. Hudson KL, Bartlett GJ, Diehl RC, Agirre J, Gallagher T, Kiessling LL, Woolfson DN (2015) J Am Chem Soc 137:15152–15160
21. Dougherty DA (2013) Acc Chem Res 46:885–893
22. Chen W, Enck S, Price JL, Powers DL, Powers ET, Wong C-H, Dyson HJ, Kelly JW (2013) J Am Chem Soc 135:9877–9884
23. Fernández-Alonso MC, Díaz D, Berbis MÁ, Marcelo F, Cañada J, Jiménez-Barbero J (2012) Curr Protein Pept Sci 13:816–830
24. Fernández-Alonso MdC, Cañada FJ, Jiménez-Barbero J, Cuevas G (2005) J Am Chem Soc 127: 7379–7386
25. Vandenbussche S, Díaz D, Fernández-Alonso MC, Pan W, Vincent SP, Cuevas G, Cañada FJ, Jiménez-Barbero J, Bartik K (2008) Chem - Eur J 14:7570–7578
26. Hansch C, Leo A, Taft RW (1991) Chem Rev 91:165–195
27. Hammett LP (1937) J Am Chem Soc 59:96–103

Part II
Carbohydrate-Modified Materials for Tissue Engineering

Chapter 4
The Modifiable Scaffold for Tissue Engineering

4.1 The Structural Basis of hSAFs

The scaffold chosen for investigating the application of carbohydrate-based modifiers for tissue engineering was the hSAF system previously developed in the Woolfson lab (Sect. 1.4.2) [1]. Hydrogels are made by mixing two synthetic peptides, hSAF-p1 and hSAF-p2 (Table 4.1), that form heterodimeric coiled coils (Fig. 1.13). These heterodimers have a designed slipped interface, leading to longitudinal assembly into fibres that entangle to form the gel (Fig. 1.12).

4.2 Peptide Synthesis and Hydrogel Preparation

For these studies, peptides were synthesised *via* automated solid-phase peptide synthesis (SPPS) using the standard fluorenylmethyloxycarbonyl (Fmoc) protecting-group strategy [2]. The typical practice for peptides designed to fold into discrete assemblies in solution is to cap the *N*-terminus with an acetyl group and have an amide at the *C*-terminus [3]. However, for the hSAF peptides the *N*-terminus is left uncapped as an amine and the *C*-terminus as a free carboxylic acid. This is crucial for successful assembly into fibres and ultimately hydrogels; it is likely that electrostatic interaction between the positively charged *N*-terminus and negative *C*-terminus (at neutral pH) contributes to longitudinal assembly. Indeed, uncapping of even blunt-ended coiled coils can lead to assembly into fibres for certain assemblies larger than dimers [4]. Crude peptides were cleaved from the solid resin support using trifluoroacetic acid (TFA) with triisopropylsilane (TIPS) as a radical scavenger and purified *via* preparative reversed-phase high-performance liquid chromatography (HPLC). The desired peptides were identified using matrix-assisted laser desorption ionization (MALDI) mass spectrometry (MS) and the purity verified by analytical HPLC.

Hydrogels are prepared from hSAF peptides by mixing hSAF-p1 and hSAF-p2 at equimolar ratio in aqueous buffer. For these studies, the final concentrations used

K. L. Hudson, *Carbohydrate-Based Interactions at the Molecular and the Cellular Level*, Springer Theses, https://doi.org/10.1007/978-3-319-77706-1_4

Table 4.1 Parent hSAF peptide sequences

Peptide	Sequence
heptad repeat	*g abcdefg abcdefg abcdefg abcdef*
hSAF-p1	H-K IAALKAK IAALKAE IAALEWE NAALEA-OH
hSAF-p2	H-K IAALKAK NAALKAE IAALEWE IAALEA-OH

for the hydrogels were the same as those originally published, i.e., at 1 mM in each peptide [1]. The peptide concentration can be varied within the range of 0.5–1.5 mM in each peptide, which alters the physical properties of the hydrogel [5], and cellular studies have been successfully carried out using gels at the lowest concentration [6]. However, the original concentration provides a good balance between gel strength and durability and conservation of material, and also gives hydrogels of the appropriate stiffness for the predominantly neural cell types being targeted [7]. Given the molecular weight of the peptides of around 3 kDa, this means that the hydrogels are >99% w/v water content.

The pH of the mixed peptide solutions is important for successful hydrogel formation, as is the procedure itself. The pH must be such that the basic Lys residues and *N*-terminus are protonated, while the acidic Glu residues and *C*-terminus are deprotonated, to allow the requisite formation of salt bridges. The gels are formed at physiological pH, 7.4, the relevant pH for mammalian cell culture and eventual application in tissue engineering, and in 3-(*N*-morpholino)propanesulfonic acid (MOPS) buffer. The peptides are initially mixed on ice, and left for 30 min before returning to room temperature. Although not essential for gel formation, this reduced temperature step does give increased gel stiffness, as to a limited extent does leaving them for up to 24 h to mature [8, 151].

4.3 Characterisation and Properties of hSAFs

Several methods can be used to characterise the hSAFs at different levels. Dilute samples (at 100 μm in each peptide, 10× lower than the concentration used for hydrogels) allow the use of techniques that require samples to be in solution, such as circular dichroism (CD) spectroscopy, as well as visualisation of the morphology of individual fibres by transmission electron microscopy (TEM). Samples of the hydrogel itself can be morphologically assessed by scanning electron microscopy (SEM), and the physical properties using rheology. The α-helical character of formed fibres, and alignment of helices along the fibre axis has previously been proven by X-ray fibre diffraction [1] and linear dichroism spectroscopy [5]; however these techniques will not be discussed further in this thesis.

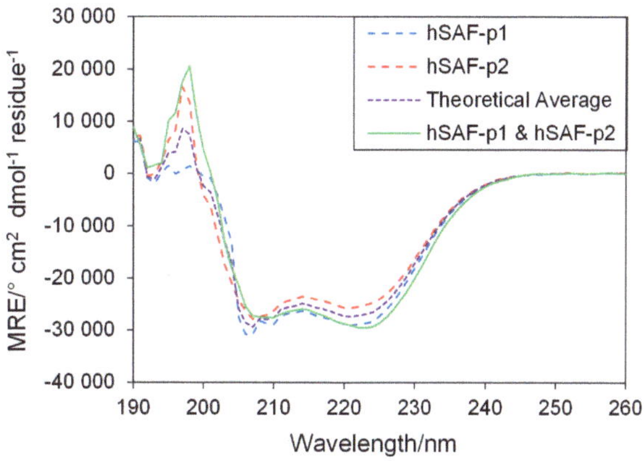

Fig. 4.1 CD spectra of peptides hSAF-p1 (blue dashed line), hSAF-p2 (red dashed line), a theoretical average of hSAF-p1 and hSAF-p2 (purple dotted line), and an equimolar mixture of hSAF-p1 and hSAF-p2 (green solid line). All at 100 μm total peptide concentration at 20 °C

Some representative data were collected for the parent hSAFs by CD spectroscopy, TEM, and rheology. These would serve as a useful reference to monitor the effects of any changes made to the peptides, and also of modification with carbohydrates.

Characterisation of Fibres by Circular Dichroism Spectroscopy

CD spectroscopy is a technique that reports on protein secondary structure by measuring the differential in absorption of oppositely circularly polarised light [9]. Peptide chains that are folded as α-helices have characteristic negative peaks of mean residual ellipticity (MRE) at 208 and 222 nm, while β-sheet structures give a single minimum MRE at 215 nm. CD spectroscopy can be used to verify hSAF formation, as the peptides are designed to be completely α-helical within the coiled-coil fibres.

The hSAF peptides are designed to form dimeric α-helical coiled coils when mixed, and also to discourage self-association. Both peptides hSAF-p1 and hSAF-p2 were also α-helical in isolation (Fig. 4.1), suggesting a degree of self-association or that they exist as single helices. When hSAF-p1 and hSAF-p2 were mixed, the interaction of the two peptides was confirmed by the fact that the spectrum for the mixture differs from the theoretical average of the spectra for the individual peptides (Fig. 4.1), and the resultant species is more folded than either peptide in isolation. Additionally, for the hSAF mixture the second minimum at 222 nm is ‘red-shifted’, so that it appears at a higher wavelength. Concomitant with this, intensity is lost at 208 nm. Both of these features are characteristic of coiled-coil-based peptide fibres, which scatter light and so alter the CD spectra [4, 10].

CD spectra were generally measured at room temperature (20 °C) and also at 37 °C, to test if there was a change in structure when the temperature was raised to that required for mammalian tissue culture. In all cases, only minor decreases in overall helicity were seen at the higher temperature, as would be expected when adding more energy to the system and so decreasing population of the most stable state. Measuring the temperature dependence of the intensity of the 222 nm minimum showed that the α-helical structure was maintained, and stable over time, up to a temperature of 45 °C, above which an irreversible transition to β-structure was observed (Sect. 7.5.1).

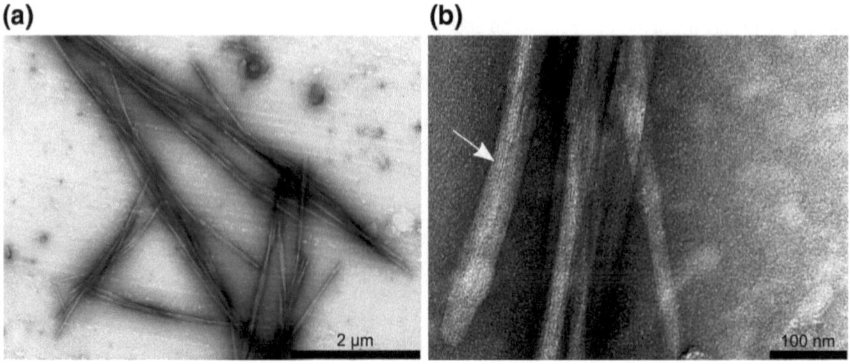

Fig. 4.2 Transmission electron micrographs of peptide fibres formed by peptides hSAF-p1 and hSAF-p2. Arrow in **b** highlights striations

Characterisation of Fibres by Transmission Electron Microscopy

hSAF fibres have diameters of 5–50 nm, and lengths of >1 μm [1]. TEM is suitable for visualising and assessing the morphology of species of this size. As for CD spectroscopy, the samples were made up at 10× dilution relative to the hydrogels to allow characterisation of individual fibres. The samples needed to be dilute to give a single layer of fibres to allow electrons to pass through the sample. The fibres were visualised by negative staining by depositing a solution containing heavy metal salt (uranyl acetate) on the sample, allowing visualisation of the sample by differentials in electron absorption [11].

TEM micrographs of hSAFs made from equimolar hSAF-p1 and hSAF-p2 (Fig. 4.2) showed fibres with the expected widths ranging from around 5–50 nm, and lengths in the order of μm. A single dimeric coiled coil has a diameter around 2 nm [12], and so the observed fibres represent the bundles of many fibrils that likely make up the hSAF hydrogels. Hints of this fine fibrillar structure were seen as longitudinal striations with widths equivalent to single fibrils at high magnification (Fig. 4.2b). These striations are characteristic of ordered peptide fibres, and have been used to determine the fine structure of coiled-coil peptide fibres, including the original SAF systems [4, 12].

Characterisation of Hydrogels by Rheology

The physical properties of the cellular environment have an impact on tissue development [7]. Some of the important physical characteristics for hydrogels can be assessed by rheology [13], e.g., the storage (elastic) modulus (G') and loss (viscous) modulus (G''). For hydrogels, G' is larger than G'' as it behaves as an elastic material and not a liquid, for which G'' is the larger. As G' describes the elastic properties, it can be used as a proxy for the stiffness of the material.

The parent hSAF hydrogels have a G' of around 1 kPa [1], which is appropriate for neural cells and brain tissue (Table 1.2) [7]. This stiffness is maintained up to a

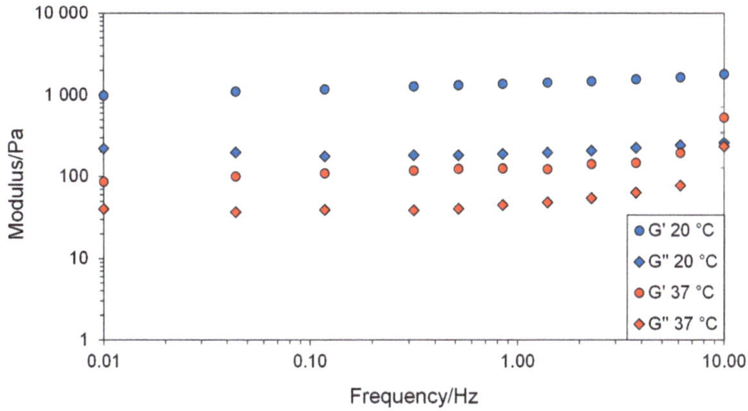

Fig. 4.3 Frequency dependence of elastic (circles) and viscous (diamonds) moduli of hydrogels made with hSAF-p1 and hSAF-p2 at 20 °C (blue) and 37 ° C (red). Error bars are standard error

temperature of 45 °C [1], above which the transition to β-structure was observed in the CD spectra.

For these studies, the rheological properties of the hSAFs were measured by forming hydrogels in situ between the rheometer plate and cone before measurement. The G′ and G″ of hSAFs showed little change with frequency (Fig. 4.3); the measurement at the lowest frequency is the most relevant, as this is closest to the resting state ('frequency = 0'). By these measurements the stiffness reduced by an order of magnitude upon heating from 20 to 37 °C, perhaps reflecting the reduced helicity observed by CD spectroscopy, but still within the appropriate range for neural cell types.

4.4 Decoration of hSAFs with Functional Motifs

A modifiable variant of the hSAFs has been developed in the Woolfson group to enhance applicability for tissue engineering by allowing tuning for specific cell types [6]. Incorporation of azide functionality into the hSAF peptides allows for modification with alkyne-functionalised motifs by CuAAC [14]. This approach has previously been used to functionalise the original SAF system with gold nanoparticles [15].

For the hSAFs, azide incorporation is achieved by substituting the *N*-terminal Lys residue of hSAF-p1 for an Anl residue (**21**), to give hSAF-p1-K1Z [6]. This is an appropriate substitution due to the structural similarity between Anl and Lys (**22**, Fig. 4.4). The position was chosen for synthetic utility; due to its relative expense, the Anl residue is manually coupled after automated SPPS before the peptide is cleaved from the resin. The *C*-terminal Ala residue can also successfully be substituted for Anl [6].

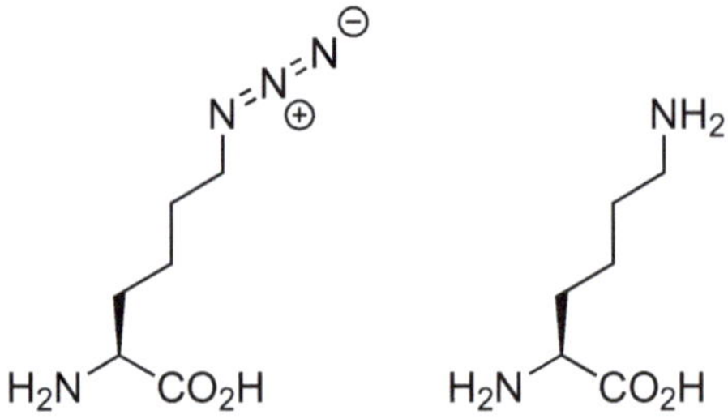

21 **22**

Fig. 4.4 Structure of Anl (**21**) and Lys (**22**)

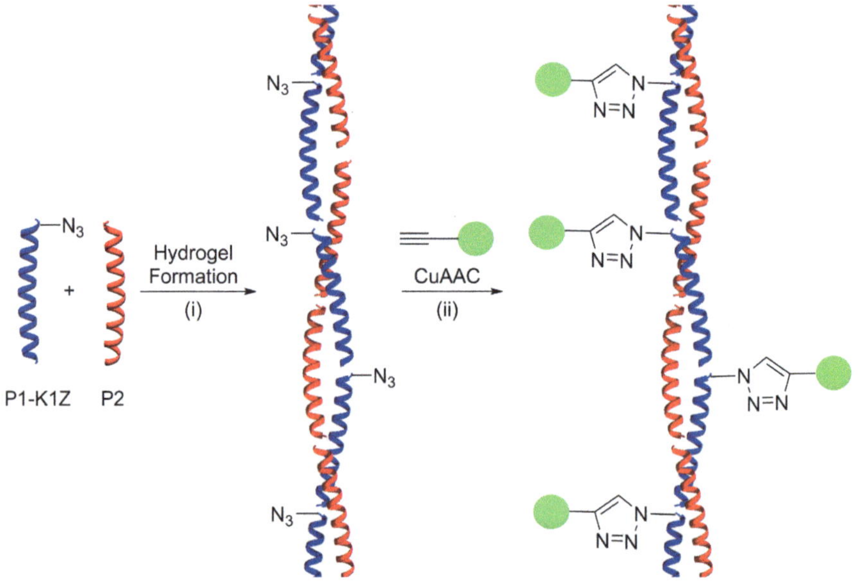

Scheme 4.1 Procedure for forming azide-containing hSAFs scaffolds and decorating with alkyne-functionalised modifier (represented as green circles). Conditions: (i) pH 7.4 aqueous (aq.) MOPS buffer, 30 min at 5 °C then room temperature (RT) overnight. (ii) 2 equiv. modifier, 4 equiv. $CuSO_4$, 4 equiv. Na ascorbate, RT overnight

hSAF-p1-K1Z can be used as a substitute for hSAF-p1 to form hydrogels with hSAF-p2. This gives an azide-functionalised hydrogel, which can subsequently be modified by CuAAC by treatment with the alkyne-functionalised modifier of interest and a solution of Cu ions (Scheme 4.1). The 'click' cycloaddition is catalysed by Cu^I ions [16], which, due to instability, are generated in situ from a solution of $CuSO_4$ and sodium ascorbate. The hydrogel is left under the modification conditions overnight to allow time for diffusion of the modifier throughout the hydrogel. After 24 h, the decorated gel is washed to remove residual copper, after which the hydrogel is ready for cell culture.

As the hydrogel fibres consist of bundles of many coiled-coil fibrils, not all azide handles will be exposed to the aqueous environment. Some azides will necessarily be buried in the hydrophobic core of the fibres, making them inaccessible to the hydrophilic modifier and coupling reagents. However, only the functional groups on the surfaces of the fibres will be accessible to cellular receptors, and so the state of any buried unreacted azides is not an issue.

Detailed cellular studies of modified hSAF scaffolds have been carried out using alkyne-functionalised peptides readily accessible by SPPS [6]. The bioactive peptide is synthesised and propiolic acid coupled to the *N*-terminus by amide coupling before cleavage from the resin. A five-amino-acid spacer is also used (GSGYG) to ensure the functional moiety is accessible to target receptors on the cell surface, and also to allow concentration determination using the ultraviolet (UV) absorbance of the Tyr moiety.

The bioactive peptide sequence RGDS was chosen for these studies (alkyne-functionalised peptide GSGYGRGDS is subsequently referred to as RGD-peptide, or RGD-0). This motif is proven to increase the cell-binding and proliferation of tissue culture scaffolds, and has become the standard for proof-of-concept studies and improving biocompatibility (Sect. 1.4.3) [17, 18]. Initial studies using PC12 cells, a model for neuronal cell lines [19], show that cell growth and proliferation is enhanced on azide-functionalised hSAF scaffolds with RGD-peptide appended to them compared to parent hydrogels [6]. The 'half-moon' model [20], where hSAF gel in a tissue culture well is functionalised on one half and the other half made up of the parent, allows direct comparison of the effect of the modifier to unmodified scaffolds. Further studies using murine embryonic neural stem cells (NSCs) have shown that RGD-decoration can encourage developmental progression towards functional neurons [21].

4.4.1 Properties of the Modifiable and Modified Scaffold

Hydrogels formed with hSAF-p1-K1Z in place of hSAF-p1 were investigated to find any differences from the parent system, and to give a baseline for the effects of decoration with carbohydrates. They had very similar properties to hSAF-p1 hydrogels, and subsequent decoration with the RGD-peptide modifier also did not alter the properties. The modification procedure makes some forms of characterisation difficult—e.g., the need to apply the reaction mixture to the surface of the hydrogel for modification is not compatible with forming gels between the rheometer plate and cone for rheology.

Circular Dichroism Spectroscopy of Modifiable and Modified Fibres

There was little difference between the CD spectra for hSAFs formed with hSAF-p1 and those formed with hSAF-p1-K1Z (Fig. 4.5). The spectrum was characteristic

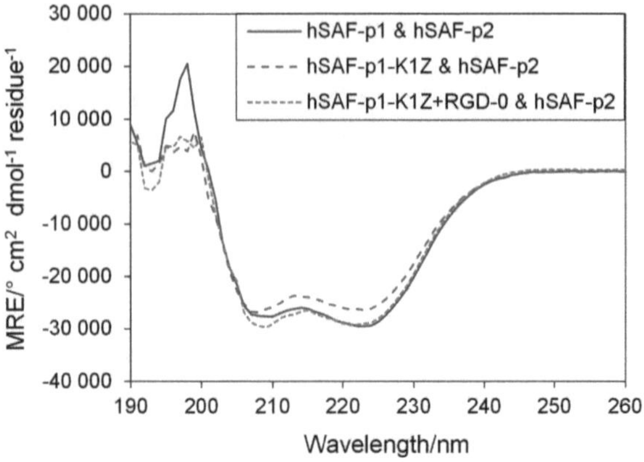

Fig. 4.5 CD spectra of fibres formed from an equimolar mixture of peptide hSAF-p2 with hSAF-p1 (blue solid line), hSAF-p1-K1Z (red dashed line), and hSAF-p1-K1Z decorated with RGD-peptide (green dotted line). All at 100 μm total hSAF peptide concentration at 20 °C

of an α-helix, showing that the substitution of Anl in place of one Lys residue does not disrupt the overall structure. There was a slight reduction in total signal at the 208 and 222 nm minima, indicating less overall folding, which may be due to the loss of the salt bridge that the Lys of the parent participates in.

Thermal unfolding of hSAF-p1-K1Z fibres again showed properties similar to the parent hSAF system, with the α-helical structure maintained to 45 °C, above which an irreversible transition to β-structure occurred (Sect. 7.5.1).

Subsequent decoration of the azide-functionalised fibres with RGD-peptide modifier also gave CD spectra characteristic of α-helices (Fig. 4.5). In fact, the total signals at the minima were the same as those of the parent hSAF system—this may be due to the bulky RGD-peptide appendage sterically disfavouring unfolded states [22].

Transmission Electron Microscopy of Modifiable and Modified Fibres

Visualisation by TEM of fibres formed from hSAF-p1-K1Z combined with hSAF-p2 (Fig. 4.6a) showed structures of similar size to those seen with combination of the parent hSAF-p1 and hSAF-p2. However, higher-magnification images revealed the presence of much smaller fibrils (Fig. 4.6b) that were not present in the samples made from the parents. These smaller fibrils were of similar lengths to the thicker fibres, but were only a few nm across and appeared to be much more flexible. It is likely that these smaller fibrils represent constituents of the larger fibres—possibly single coiled-coil fibrils—that have not bundled. It is not clear why they were not seen for the parent peptides; this may be due to changes in the dynamics of fibre formation.

The images of the smaller fibrils were reminiscent of those seen in a published study of the constituents of peptide fibres [23], including repeating light and dark bands. These were possibly caused by turns within the fibre—the super-coiling of the coiled coil means that there will be periodic changes in topology of fibres and fibrils as the peptides wrap around each other (the superhelical pitch). A study of the original SAFs showed a superhelical pitch of 12.5 nm, approximately what was observed for the hSAF fibrils. Interestingly, several natural 3D matrices formed by protein fibres such as collagen consist of large bundles of fibres enmeshed within

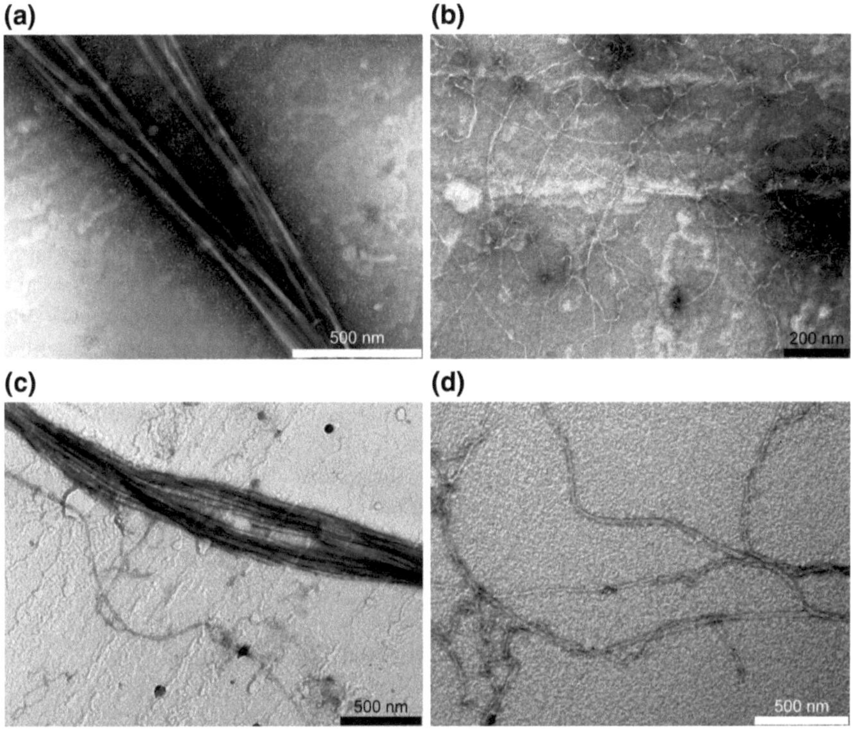

Fig. 4.6 Transmission electron micrographs of peptide fibres formed by peptides hSAF-p1-K1Z and hSAF-p2 **a, b** unmodified and **c, d** modified with RGD-peptide

much smaller individual fibrils [24]. This may be the form that the hSAF scaffolds take, given that they are kinetically trapped, but visualisation of the smaller fibrils at gelation concentrations is challenging.

Smaller fibrils were also apparent alongside larger fibres for samples of fibres decorated with RGD-peptide prior to visualisation (Fig. 4.6c, d). This shows that the formation of the fibres before decoration dominates the morphology, and is not altered by the decoration process.

4.4.2 Assessment of Copper Content of Modified Scaffolds

When using CuAAC to modify scaffolds for tissue culture, it is important to ensure that no potentially toxic copper remains that could adversely affect the cells [25, 26]. To remove residual copper after the CuAAC step, the decorated gels were washed with aqueous ethylenediaminetetraacetic acid (EDTA), and then further washing with H_2O or the relevant buffer was carried out to remove EDTA and any other reagents.

The survival of cells on decorated hSAF scaffolds has previously been taken as an indication that the copper content of the hydrogels is below fatal levels, but it was not known exactly how much remained in the gel and if this affected cellular development.

Perhaps the most accurate way to assess the levels of copper in a sample is by atomic absorption spectroscopy. However, this requires relatively large samples (10 mL) to give a sensitivity to around 1 mg L^{-1}, which is the level that the sample needs to be below to ensure non-toxicity. This sample size requirement is too large for repeated measurements of the current hSAF system, and dilution of smaller samples would reduce the sensitivity below useful levels. Instead, an assay using UV-Vis spectrometry was used. Bicinchoninic acid (BCA) forms a complex with Cu^{I} with a characteristic absorption of light at 355 nm in the UV band [27]. A calibration curve was obtained and assays were carried out on samples that had been washed following the procedures originally used (wash with EDTA 10× followed by buffer 10×), and a reduced washing procedure with each only three times. Both showed only baseline adsorption at 355 nm, indicating levels of Cu below 1 mg L^{-1}. All subsequent studies were completed using the reduced washing.

4.5 Discrete Heterodimers for Method Development

Understanding the nature of the ligation process of modifiers onto the peptides was a key aspect of this research. Characterisation and monitoring is limited for the hydrogels, often requiring disruption of the gel itself; both are much easier for samples in solution with more techniques available. Reactions on free, unpaired hSAF-p1-K1Z are not representative of those in the hydrogel as the peptide is not in the coiled-coil conformation and many of the amino-acid residues are not involved in their designed interactions. The dilute fibres investigated in Sect. 4.4 are not truly in solution. A system with all the residues involved in their intended interactions but the species in solution was thus developed to allow optimisation of the decoration procedure.

A peptide was designed to form a more-traditional blunt-ended heterodimeric coiled coil with hSAF-p1 peptides. This was based on hSAF-p2, but with the first and second halves swapped around while maintaining parallel assembly, so that pairing with hSAF-p1 leaves no 'sticky ends', thereby preventing longitudinal assembly (Fig. 4.7). This approach has previously been undertaken with the original SAF system [10]. The designed peptide was then compatible with any modified hSAF-p1 sequence intended to be used with the hSAF system, to allow verification that the alterations do not disrupt the coiled coil formation.

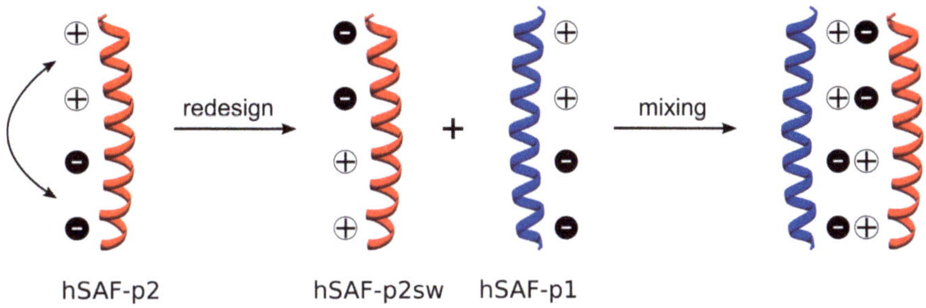

Fig. 4.7 Design concept for discrete heterodimeric coiled coils from hSAF peptides: swapping the first and second halves of hSAF-p2, while maintaining parallel assembly, prevents formation of sticky ends and instead leads to formation of blunt-ended heterodimers

Fig. 4.8 CD spectra of peptides hSAF-p1 (blue dashed line), hSAF-p2sw (red dashed line), a theoretical average of hSAF-p1 and hSAF-p2sw (purple dotted line), and an equimolar mixture of hSAF-p1 and hSAF-p2sw (solid green line). All at 100 μm total peptide concentration at 20 °C

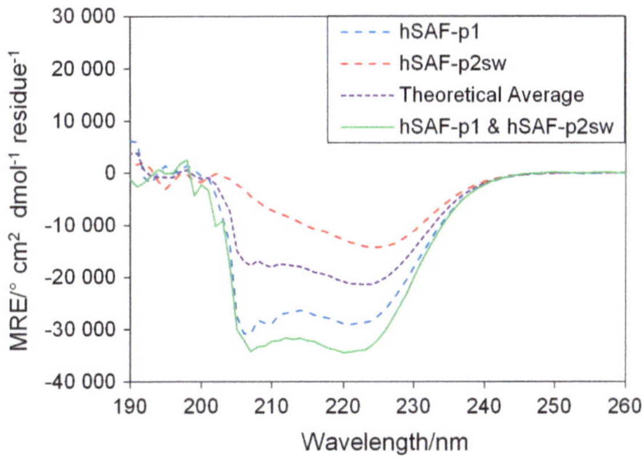

4.5.1 *Properties of Discrete Heterodimers*

CD spectroscopy of hSAF-p2sw with hSAF-p1 showed that together they form an α-helical coiled coil as expected (Fig. 4.8). Interestingly, simply swapping the first and second halves made hSAF-p2sw much less helical in isolation compared to hSAF-p2 (Fig. 4.1). This is surprising, as for single peptides α-helicity should be more favoured when the side chains at the termini are of opposite charge to the terminus (as for hSAF-p2sw, with Glu residues near the *N*-terminus and Lys residues near the *C*-terminus) [28]. hSAF-p2sw did however clearly interact with the complementary peptide hSAF-p1 as the mixture was comparable in level of α-helicity to the corresponding fibres (Fig. 4.1), without the red-shifting of the 222 nm minimum that is characteristic of fibre formation.

To confirm that the assemblies formed were dimers and not larger assemblies the species weight of equimolar hSAF-p1 and hSAF-p2sw was assessed by analytical ultracentrifugation (AUC) [29], and the data fitted to a single ideal species with the mass expected for a heterodimer (see Sect. 7.5.2).

As expected, hSAF-p1-K1Z also gave a CD spectrum when mixed with hSAF-p2sw equivalent to that seen when it was mixed with hSAF-p2.

With the modifiable scaffold in place, along with these tools to determine and analyse the effects of functionalisation with different motifs, I was able to develop appropriate carbohydrate modifiers for use with the system. These were alkyne-functionalised to allow decoration of the azide-containing scaffolds *via* CuAAC. A variety were made for proof-of-principle experiments, as well as those to investigate the effects of linker lengths and oligosaccharides for application to targeted tissue types.

References

1. Banwell EF, Abelardo ES, Adams DJ, Birchall MA, Corrigan A, Donald AM, Kirkland M, Serpell LC, Butler MF, Woolfson DN (2009) Nat Mater 8:596–600
2. Fields GB, Noble RL (1990) Int J Pept Protein Res 35:161–214
3. Fletcher JM, Boyle AL, Bruning M, Bartlett GJ, Vincent TL, Zaccai NR, Armstrong CT, Bromley EHC, Booth PJ, Brady RL, Thomson AR, Woolfson DN (2012) ACS Synth Biol 1:240–250
4. Burgess NC, Sharp TH, Thomas F, Wood CW, Thomson AR, Zaccai N, Brady RL, Serpell LC, Woolfson DN (2015) J Am Chem Soc 137:10554–10562
5. Wasmuth A, Mehrban N, Hudson KL, Thomson AR, Mullen LM, Birchall MA, Woolfson DN Manuscript in preparation
6. Mehrban N, Abelardo ES, Wasmuth A, Hudson KL, Mullen LM, Thomson AR, Birchall MA, Woolfson DN (2014) Adv Healthc Mater 3:1387–1391
7. Discher DE, Mooney DJ, Zandstra PW (2009) Science 324:1673–1677
8. Bromley EHC, Channon KJ, King PJS, Mahmoud ZN, Banwell EF, Butler MF, Crump MP, Dafforn TR, Hicks MR, Hirst JD, Rodger A, Woolfson DN (2010) Biophys J 98:1668–1676
9. Chen Y-H, Yang JT, Chau KH (1974) Biochemistry 13:3350–3359
10. Pandya MJ, Spooner GM, Sunde M, Thorpe JR, Rodger A, Woolfson DN (2000) Biochemistry 39:8728–8734
11. De Carlo S, Harris JR (2011) Micron 42:117–131
12. Sharp TH, Bruning M, Mantell J, Sessions RB, Thomson AR, Zaccai NR, Brady RL, Verkade P, Woolfson DN (2012) Proc Natl Acad Sci USA 109:13266–13271
13. Yan C, Pochan DJ (2010) Chem Soc Rev 39:3528–3540
14. Rostovtsev VV, Green LG, Fokin VV, Sharpless KB (2002) Angew Chem Int Ed 41:2596–2599
15. Mahmoud ZN, Gunnoo SB, Thomson AR, Fletcher JM, Woolfson DN (2011) Biomaterials 32:3712–3720
16. Worrell BT, Malik JA, Fokin VV (2013) Science 340:457–460
17. Hersel U, Dahmen C, Kessler H (2003) Biomaterials 24:4385–4415
18. Perlin L, MacNeil S, Rimmer S (2008) Soft Matter 4:2331–2349
19. Greene LA, Tischler AS (1976) Proc Natl Acad Sci USA 73:2424–2428
20. Chan TR, Stahl PJ, Yu SM (2011) Adv Funct Mater 21:4252–4262
21. Mehrban N, Zhu B, Tamagnini F, Young FI, Wasmuth A, Hudson KL, Thomson AR, Birchall MA, Randall AD, Song B, Woolfson DN (2015) ACS Biomater Sci Eng 1:431–439
22. Price JL, Powers ET, Kelly JW (2011) ACS Chem Biol 6:1188–1192
23. Aggeli A, Nyrkova IA, Bell M, Harding R, Carrick L, Mcleish TCB, Semenov AN, Boden N (2001) Proc Natl Acad Sci USA 98:11857–11862
24. Ushiki T (2002) Arch Hist Cytol 62:109–126
25. Gaetke LM, Chow CK (2003) Toxicology 189:147–163

26. Schilsky ML, Blank RR, Czaja MJ, Zern MA, Scheinberg IH, Stockert RJ, Sternlieb I (1989) J Clin Invest 84:1562–1568
27. Brenner AJ, Harris ED (1995) Anal Biochem 226:80–84
28. Baker EG, Bartlett GJ, Crump MP, Sessions RB, Linden N, Faul CFJ, Woolfson DN (2015) Nat Chem Biol 11:221–228
29. Schuck P (2000) Biophys J 78:1606–1619

Chapter 5
Carbohydrate Modifiers for Tissue Engineering Scaffolds

5.1 Synthesis of Simple Carbohydrate Modifiers

Three monosaccharides, Glc (**23**), Gal (**24**), and GlcNAc (**25**, Fig. 1.2) were selected as initial targets for the synthesis of carbohydrate modifiers for the hSAF scaffold. These were chosen due to their prevalence in biology as some of the major components of natural oligosaccharides [1], meaning that they are most likely to be tolerated by cells and elicit a biological response. Their ubiquity also means that they are readily available and accessible syntheses of many derivatives have been developed.

Gal has applications in liver tissue engineering [2], and is also implicated in oligodendrocyte differentiation, along with the sulfated derivative SGal (**15**) [3], making it potentially useful for neural tissue engineering (Sect. 1.4.4). GlcNAc is of interest as it is the monosaccharide in natural *N*-glycans attached covalently to Asn residues to form glycoproteins upon which further units are appended to construct glycans [4, 5]. In addition, single GlcNAc residues are also often appended to proteins in nature through *O*-linked glycosylation, and it is a monomer component of ECM polysaccharides such as hyaluronan [6]. Glc provides an appropriate control as it rarely features in functional glycans [1], probably due to its ubiquity as a metabolite.

The simplest way to functionalise these monosaccharides with an alkyne moiety was as propargyl glycosides. The anomeric position is appropriate for modification of carbohydrates as it is the most-commonly substituted site in natural oligosaccharides, and so has the least chance of disrupting biological activity. Numerous examples of propargyl glycosides have been synthesised by various methods [7, 8], including for the purpose of linkage of the sugars to polypeptides *via* CuAAC [9].

Alkyne-functionalised derivatives of Glc (**26**), Gal (**27**), and GlcNAc (**28**) were synthesised (Fig. 5.1). This gave tools to append to the azide-funtionalised hSAFs by CuAAC to verify the decoration and investigate the effect on the scaffold properties, and possible responses by cells. Propargyl SGal (**29**), synthesised by an adapted method [10], was kindly provided for studies on oligodendrocytes by Prof. Amit Basu (Brown University, Providence, RI, USA).

K. L. Hudson, *Carbohydrate-Based Interactions at the Molecular and the Cellular Level*, Springer Theses, https://doi.org/10.1007/978-3-319-77706-1_5

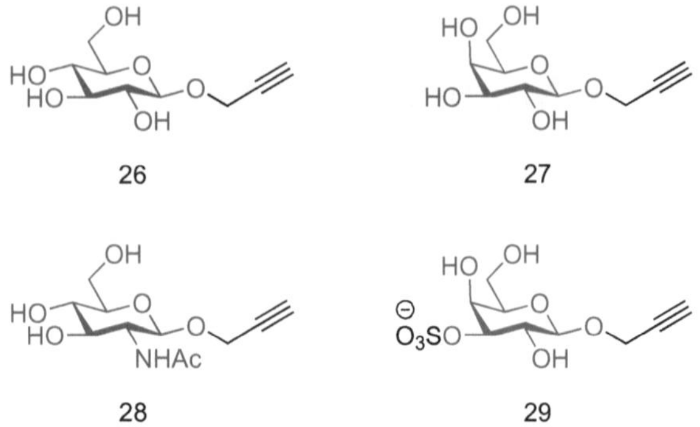

Fig. 5.1 *O*-Propargyl glycopyranosides of Glc (**26**), Gal (**27**), GlcNAc (**28**), and SGal (**29**)

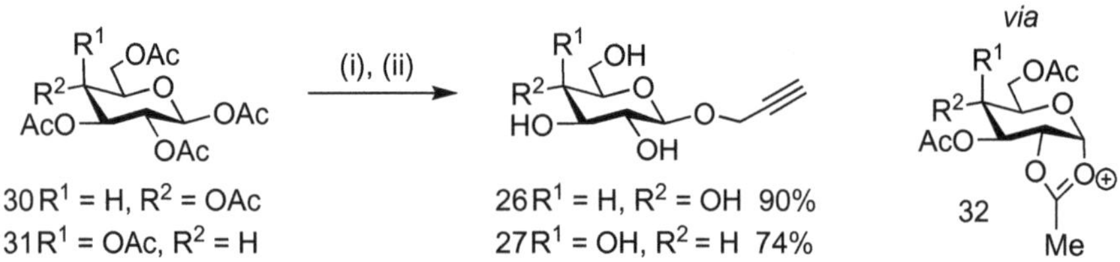

Scheme 5.1 Synthesis of *O*-propargyl gluco- (**26**) and galactopyranosides (**27**). Conditions: (i) 2-Propyn-1-ol (1.2 equiv.), $BF_3.OEt_2$ (1.5 equiv.), anhydrous anhydr. CH_2Cl_2, RT, 2 h. (ii) NaOMe (20 equiv.), (anhydr) MeOH, RT, 2 h

5.1.1 Synthesis of Propargyl Glc, Gal, and GlcNAc

O-Propargyl Glc (**26**) and Gal (**27**) were synthesised by a published method [7]. This involved conversion of peracetylated pyranoses **30** and **31** into *O*-propargyl tetra-acetylated pyranoside by glycosylation of propargyl alcohol, with boron trifluoride acting as a Lewis acid (Scheme 5.1).

One particular anomer was not targeted for the products, as most monosaccharides occur naturally as both α- and β-anomers. However, the method used was only successful with the β-anomer of the starting material, giving the β-product exclusively. This is due to the reaction proceeding *via* a 5-membered dioxalenium (**32**, Scheme 5.1) formed through neighbouring-group participation of the O(2)-acetyl group [11]. This requires donation of lone pair electrons of the acteyl oxygen into the C–O σ* orbital to break the covalent bond at the anomeric position, which is only possible with the β-anomer. Therefore, these reactions were carried out using commercially available peracetylated β-carbohydrates as starting materials. Peracetylation of the unprotected monosaccharide is possible, but gives a mixture mostly consisting of the unreactive α-anomer, due to the anomeric effect. The α-product can be accessed by a Fischer glycosylation of propargyl alcohol [12].

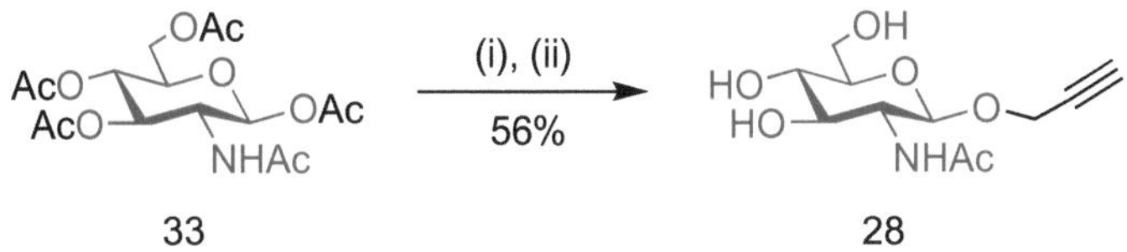

Scheme 5.2 Synthesis of *O*-propargyl *N*-acetylglucopyranoside. Conditions: (i) 2-Propyn-1-ol (20 equiv.), TMSOTf (1.1 equiv.), anhydr CH_2Cl_2, RT, 24 h. (ii) catalytic (cat.) NaOMe, anhydr MeOH, RT, 2 h

Successful coupling of deprotected *O*-propargyl sugars with proteins by CuAAC in water has previously been reported [9], and so the tetraacetylated pyranosides of Glc and Gal were fully deacetylated to give **26** and **27** for ligation to the hSAFs. This was achieved using catalytic sodium methoxide in methanol (Scheme 5.1), with the reaction occurring almost instantaneously in quantitative yield.

The importance of the 2-acetyl group in the mechanism of substitution at the anomeric position was emphasised by the fact that the analogous procedure for *N*-acetylglucosamine (**25**) required use of a different Lewis acid than boron trifluoride. This is due to the presence of an amide group at the 2-position of the peracetylated derivative rather than an ester group. This means that the reaction proceeds *via* an oxazoline rather than a dioxalenium, which is more stable, and so requires a stronger activating group to enable reaction with the alcohol. TMS triflate was used to promote the glycosylation of peracetylated GlcNAc (**33**) by a published procedure [13], giving the desired *O*-propargyl tetraacetylated pyranoside, followed by deprotection with NaOMe, as used previously, to give propargyl GlcNAc (**28**, Scheme 5.2).

5.2 Properties of Carbohydrate-Modified hSAFs

With the simple alkyne-functionalised carbohydrate modifiers Glc-0-alkyne[1] (**26**), Gal-0-alkyne (**27**), and GlcNAc-0-alkyne (**28**) in hand, the next step was to use them to decorate hSAFs and determine any effects on the properties. This was achieved by using the same fibre and hydrogel formation and CuAAC conditions as for the peptide modifiers such as RGD-peptide.

The propargyl glycosides were first reacted with free hSAF-p1-K1Z in solution to investigate whether the unprotected alkynyl modifiers could be appended onto azide-functionalised peptides by CuAAC. The successful ligation was verified by analytical HPLC. The carbohydrates are relatively polar, water-soluble moieties, and so modification of the peptide with each of the glycosides reduced the retention time of the peptide in reversed-phase HPLC, with a gradient of increasing MeCN in H_2O (Fig. 5.2). Successful addition of the carbohydrates to the peptide was verified by MALDI MS, showing completed reaction in all cases within 30 min. These tech-

[1] See Abbreviations and Acronyms for full explanation of this notation.

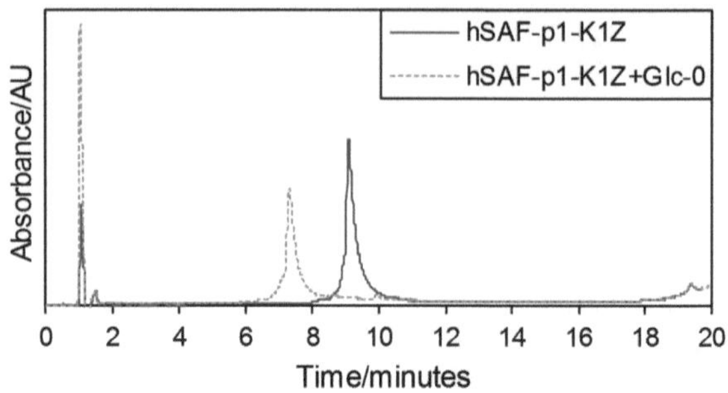

Fig. 5.2 Analytical HPLC chromatograms for hSAF-p1-K1Z undecorated (blue solid line) and decorated with Glc-0-alkyne (green dashed line) by CuAAC. Conditions: 30–50% MeCN in H_2O over 15 min

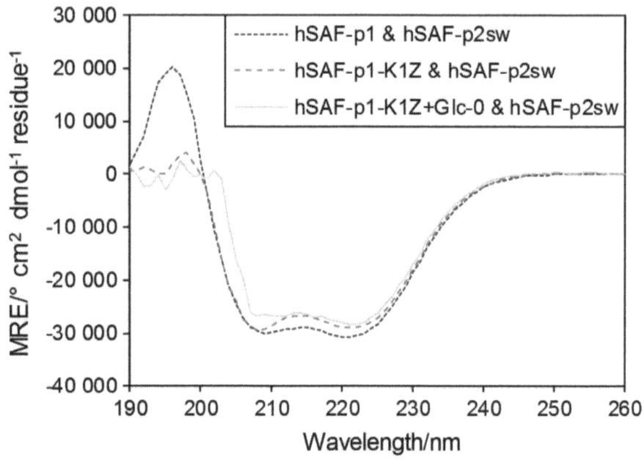

Fig. 5.3 CD spectra of peptide hSAF-p2sw mixed with hSAF-p1 (blue dotted line), hSAF-p1-K1Z (red dashed line), and hSAF-p1-K1Z decorated with Glc-0-alkyne (green solid line). All at 100 μM total peptide concentration at 20 °C

niques were used to verify glycosylation of the hSAF-p1-K1Z component of fibres and hydrogels in subsequent studies.

CD spectroscopy showed slightly reduced α-helicity for hSAF-p1-K1Z modified with Glc-0-alkyne compared to undecorated hSAF-p1-K1Z. However, when dimerised with hSAF-p2sw the overall helicity of the mixture was almost identical to that of the undecorated dimer (Fig. 5.3). This was still slightly less than for the parent hSAF system, but this likely reflects the fact that after modification the hSAF-p1-K1Z peptide again had one fewer Lys residue available to form inter-helix electrostatic interactions compared to hSAF-p1.

Fibres formed with hSAF-p2 from hSAF-p1-K1Z modified with Glc-0-alkyne had a similar CD spectrum to those formed with unmodified hSAF-p1-K1Z (Fig. 5.4). The Glc-modified fibres did not have increased helicity matching that of the parents as was seen for those modified with RGD-peptide (see Fig. 4.5). This may be due to the reduced steric bulk of the Glc compared to RGD-peptide, meaning that the monosaccharide has less effect on the secondary structure.

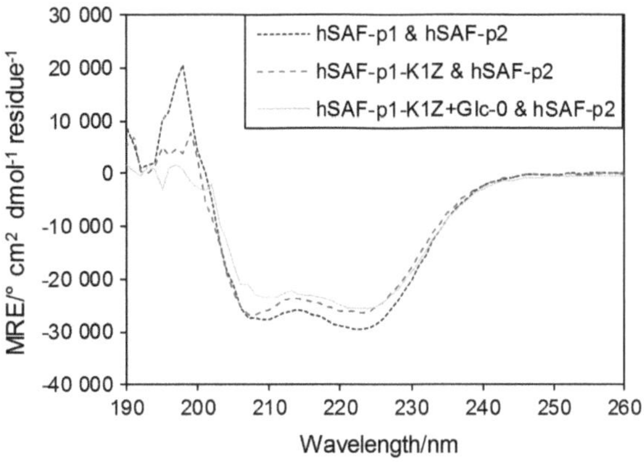

Fig. 5.4 CD spectra of peptide hSAF-p2 mixed with hSAF-p1 (blue dotted line), hSAF-p1-K1Z (red dashed line), and hSAF-p1-K1Z decorated with Glc-0-alkyne (green solid line). All at 100 μM total peptide concentration at 20 °C

Visualisation of fibres formed from hSAF-p1-K1Z modified with the propargyl glycosides by TEM showed species of similar morphology to those formed with undecorated hSAF-p1-K1Z, and also those formed from hSAF-p1-K1Z modified with RGD-peptide. Both the larger bundled fibres and the individual fibrils were present (Sect. 7.5.1).

5.3 Cellular Response to Glycosylated hSAFs

Given the results showing that hSAF can be modified with the simple propargyl glycosides, the viability of cells grown on the glycosylated scaffolds was assessed. With only the monosaccharides appended, it was not expected that any specific cellular response would be observed. Instead, these experiments were to determine the compatibility of the carbohydrate-modified scaffolds with mammalian cell culture. Murine fibroblast (3T3) cells are suitable mammalian cells for such proof-of-concept studies due to their robustness. This is because of their role in wound healing, which also makes them relevant for tissue-engineering applications.

Samples of hSAF-p1-K1Z scaffolds decorated with Glc-0-alkyne, Gal-0-alkyne, and GlcNAc-0-alkyne were used to test qualitatively whether cells cultured on glycosylated scaffolds remained viable, along with negative controls of hydrogels formed with hSAF-p1 and undecorated hSAF-p1-K1Z. Positive controls were hSAF-p1-K1Z hydrogel decorated with RGD-peptide, BME, and TCP. Cells were seeded on the various substrates and incubated at 37 °C for 3 d, and then live and dead cells were visualised by fluorescence microscopy. Two fluorescent dyes were used: calcein, a fluoroscein derivative that visualises live cells as it is actively transported across membranes; and propidium iodide, which visualises dead cells as it can only traverse the permeable membranes of dead cells, where it binds to DNA.

Upon visualisation after the 3d incubation, no major differences in densities of live cells on each of the samples were observed by visual inspection, including

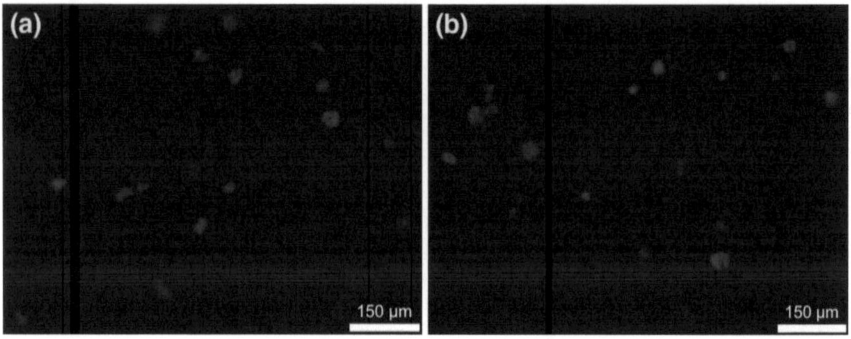

Fig. 5.5 Fluorescence micrographs of cells seeded on gels made with **a** hSAF-p1 and hSAF-p2, and **b** hSAF-p1-K1Z and hSAF-p2 decorated with Glc-0-alkyne. Samples were incubated for 3 d and stained with calcein

the undecorated gels and all of the carbohydrate-modified examples (Fig. 5.5 and Sect. 7.7.1). Some limited cell death is always expected during cell culture, and the dead staining showed dead cells on the unmodified hydrogels, BME, and the gel modified with RGD-peptide. Interestingly, however, no dead cells were seen on any of the three scaffolds decorated with the different monosaccharides.

5.3.1 Cellular Proliferation on Carbohydrate-Modified Scaffolds

After the successful viability experiments, the effect of the three monosaccharides on cell proliferation was investigated by metabolic assays on cells cultured on the modified scaffolds. The assay measures proliferation by the addition of the tetrazolium dye MTT, which is reduced to insoluble formazan with a characteristic purple colour only by viable cells [14]. The absorbance of visible light after addition of MTT can therefore be used to quantify the number of viable cells present. Monitoring this over a period of time is indicative of cellular proliferation. A 14 d study was carried out, with undecorated parent hSAF as well as hSAF-p1-K1Z gels modified with Glc-0-alkyne, Gal-0-alkyne, and GlcNAc-0-alkyne. Surprisingly, this showed not only equivalent but increased proliferation of the cells on the hSAF scaffolds that were modified with all three monosaccharides compared to the undecorated parent hSAF after 7 d and particularly 14 d (Fig. 5.6). While the absorbance correlating to number of living cells stayed approximately constant for the unmodified hSAF, it was more than double after 14 d for each of the samples modified with carbohydrates

These results suggest that carbohydrate-modified scaffolds provide a more-favourable environment for mammalian cell culture, and are in accordance with the preliminary viability studies that showed no dead cells on the carbohydrate-modifed samples. There are several possible explanations for this increased proliferation: it

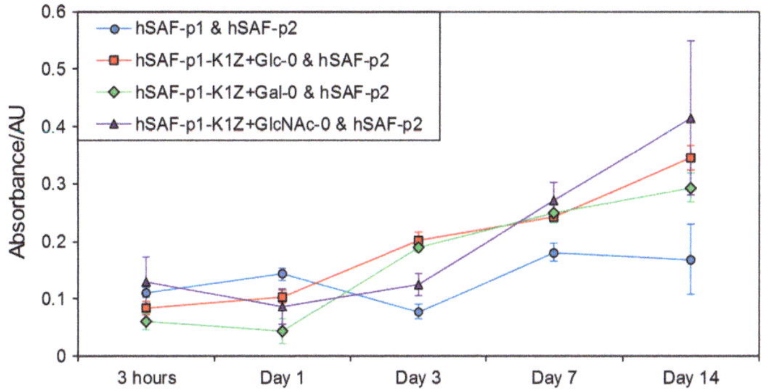

Fig. 5.6 14 d cell proliferation study for murine fibroblasts cultured on hSAF scaffolds made with hSAF-p1 (blue) and hSAF-p1-K1Z modified with Glc-0-alkyne (red), Gal-0-alkyne (green), and GlcNAc-0-alkyne (purple). Absorbance of samples at 560 nm after MTT assay (correlates to number of live cells) at indicated time points. Error bars are standard error.

could be due to recognition of the monosaccharides as non-specific biological cues by the cells; it could also be due to increased biocompatibility of the scaffolds from their increased hydrophilicity. Similar increases in cell proliferation have been seen in studies on glycosylated organic polymers [15]. It is also possible that the glycosylation specifically, or the modification more generally, favourably alters the physical properties of the hydrogel.

It is interesting that the most successful carbohydrate after 14 d was apparently GlcNAc (although standard error is large for this experiment) given that this is the only of the three investigated that is a component of a polysaccharide found naturally in the ECM. Also, GlcNAc is often the monosaccharide appended directly onto proteins in nature, i.e., if a protein is glycosylated with only a monosaccharide in nature, it is almost always with GlcNAc.

5.4 Decoration of hSAFs Before Hydrogel Formation

The first modifiable hSAF system requires the decoration of the hydrogel by addition of the modifier and reagents for CuAAC after formation of the hydrogel [16, 17]. An alternative system, where modifications are made to the azide-containing peptide by CuAAC before hydrogel formation would have a number of advantages. There would be: no need to leave hydrogel decorating overnight; no need for repeated washing to remove copper from hydrogel, and no chance of residual copper in the hydrogel; no potential change in hydrogel properties during exposure to decoration conditions; easy control of composition of modifiers in final hydrogel; ability to measure directly the rheological properties of modified hydrogel.

5.4.1 Modification of hSAF-p1-K1Z Before Hydrogel Formation

One possible approach to modified hydrogels would be to pre-decorate the existing peptide hSAF-p1-K1Z before combination with hSAF-p2 to try and form fibres or hydrogels directly. However, previous attempts to achieve this using hSAF-p1-K1Z decorated with RGD-peptide failed [16]. This may have been due to the additional bulky motif at the *N*-terminus of the peptide preventing the key non-covalent interactions that drive heterodimer or longitudinal assembly, or bundling of fibrils. The Glc-0-alkyne motif is much smaller than the RGD-peptide, and so despite this past failure I considered that addition of the monosaccharide would still allow fibres to form.

hSAF-p1-K1Z was modified with Glc-0-alkyne and then purified by HPLC. However, when combined with hSAF-p2 under hydrogelation concentrations no gel formation was seen. This was the case on ice, at room temperature, or upon leaving the sample overnight. Therefore it seems that predecoration is not possible with hSAF-p1-K1Z peptide, even with the minimal Glc-0 group.

The morphology of any fibres formed using the hSAF-p1-K1Z pre-modified with Glc-0-alkyne was investigated to try to understand why gelation was not successful. When hSAF-p1-K1Z ligated with Glc-0-alkyne was combined with hSAF-p2, TEM revealed large fibres consistent with all of the previous samples (Fig. 5.7a). However, smaller fibrils seen with other hSAF-p1-K1Z preparations, including those modified after fibre formation, were not present. Instead much shorter and stiffer fragments that stained much more darkly were observed (Fig. 5.7b). This supports the idea that the decoration may impede longitudinal assembly of the fibres, and also suggests that the small fibrils may be key to hydrogel assembly.

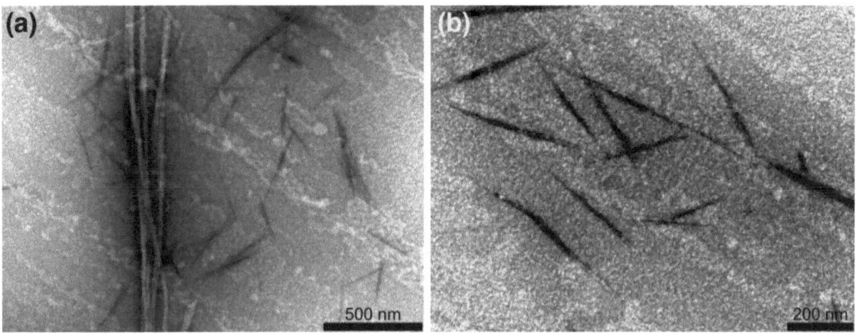

Fig. 5.7 Transmission electron micrographs of fibres formed by mixing peptides hSAF-p1 pre-modified with RGD-peptide and hSAF-p2

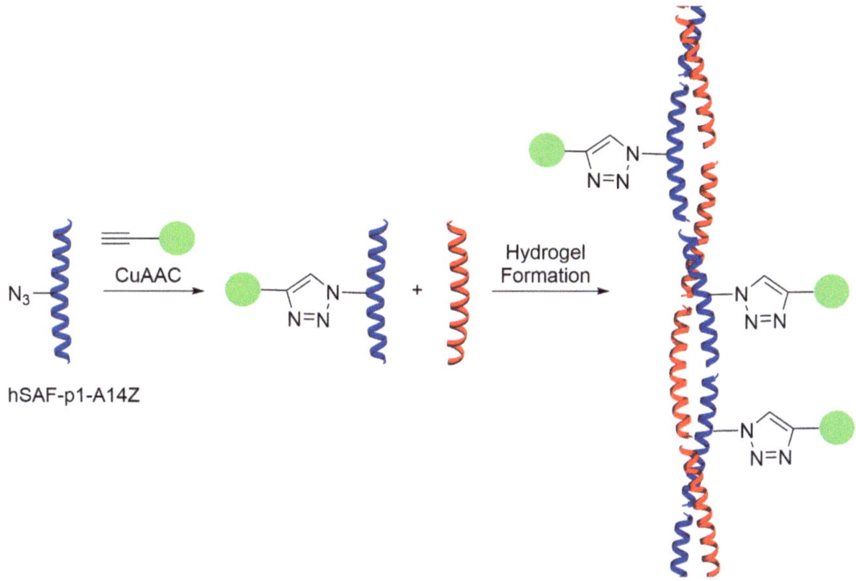

Fig. 5.8 Route to decorated hydrogels by modification of azide-containing peptide in solution before hydrogel formation. Substituting Anl for Ala at position 14, an *f*-position (away from the dimer interface) at the middle of the peptide allows for successful hydrogel formation after CuAAC

5.4.2 Modification of hSAF-p1-A14Z Before Hydrogel Formation

Disruption of inter-helix electrostatic interactions, either across the heterodimer interface or longitudinally at the peptide termini, is one of the most likely reasons that modification of free hSAF-p1-K1Z prevents hydrogel formation. Given this, an alternative site to the K1 position for substitution of Anl was sought. In dimeric coiled coils, the *f*-position is located furthest from the dimer interface (Fig. 1.13a), and so modification here is least likely to disrupt dimerisation. In the hSAF peptides, A14 is the most central *f*-position (Table 4.1), and so I argued that modification at this position would be least likely to affect longitudinal fibril assembly. Therefore, a variant of hSAF-p1 was synthesised with Anl replacing A14, hSAF-p1-A14Z (Fig. 5.8).

Substitution of Anl at a central *f*-position has been tried with the SAFs, and prevents successful formation of fibres, instead giving β-structure [18], but given the differences in the hSAF peptide sequences I hoped this would not occur in this case. Continuation of automated SPPS after manual coupling of Anl proceeded without issue.

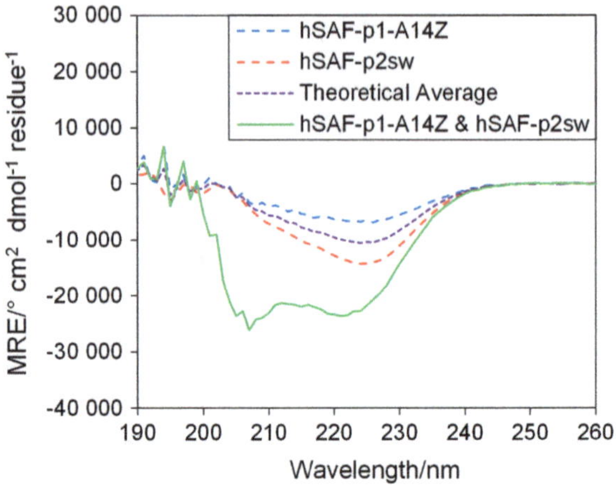

Fig. 5.9 CD spectra of hSAF-p1-A14Z and hSAF-p2sw, with a theoretical average of the two, and of an equimolar mixture of the two. All at 100 μM total hSAF peptide concentration at 20 °C

Characterisation and Properties of hSAF-p1-A14Z Fibres and Hydrogels

After synthesis and purification, it was investigated whether substituting Anl for Ala in the middle of the sequence changed the key properties of the hSAF.

CD spectroscopy of hSAF-p1-A14Z showed that it was much less folded as an α-helix than either isolated hSAF-p1 or hSAF-p1-K1Z (Fig. 5.9), so the substitution at position 14 did alter the structure of the free peptide. However, it is the interaction of the peptide with the complementary hSAF peptides that is most relevant. When hSAF-p1-A14Z was mixed with hSAF-p2sw in solution (Fig. 5.9), it interacted to give a species with total α-helicity comparable to that of hSAF-p1-K1Z with hSAF-p2sw (Fig. 5.3), indicative of the formation of heterodimeric coiled coils. CD spectroscopy of the hSAF-p1-A14Z combined with hSAF-p2 also showed formation of a helical species

Mixing hSAF-p1-A14Z with hSAF-p2 under conditions to form hydrogels was successful, confirming that changing the position of the azide did not disrupt this key feature. Rheology of the hSAF-p1-A14Z hydrogels showed that the gel was much less stiff than the parent gel at room temperature (Fig. 5.10), although it was still robust enough to handle. However, in contrast to the parent system, the hydrogels increased in stiffness at the higher temperature of 37 °C. At this temperature, which is more relevant for cell culture, the gels are of around the appropriate stiffness for neural tissue engineering (Table 1.2).

hSAF-p1-A14Z was modified with the various simple alkyne-containing carbohydrate modifiers by the same procedure previously used for free peptides in solution (Sect. 4.4), and the adducts were purified by HPLC. The resultant glycosylated peptides showed no absorbance characteristic of α or β secondary structure. Glycosylation seems to destabilise the little α-helicity seen for undecorated hSAF-p1-A14Z. Mixing with hSAF-p2 again led to CD spectra characteristic of α-helices (Sect. 7.5.1).

Hydrogels formed from hSAF-p1-A14Z decorated with Glc-0-alkyne and Gal-0-akyne had G′ values of the same order of magnitude as the parent hSAFs (Fig. 5.10),

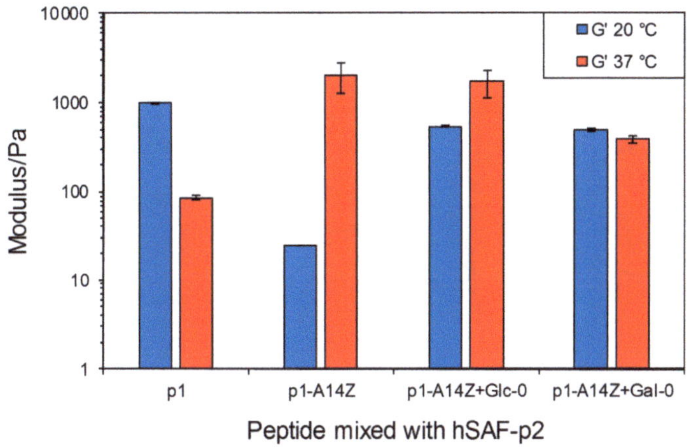

Fig. 5.10 G′ values for hydrogels formed with hSAF-p1, hSAF-p1-A14Z, and hSAF-p1-A14Z pre-modified with Glc-0-alkyne and Gal-0-alkyne at 20 °C (blue) and 37 °C (red). Values obtained at 0.1 Hz frequency, i.e., that most representative of the resting state. Error bars are standard error

showing that appendage of the carbohydrate moieties does not prevent gel formation. Both formed much stiffer hydrogels at room temperature than undecorated hSAF-p1-A14Z, which suggests that the azide may be disrupting assembly at this temperature. The two carbohydrates led to slightly different responses to temperature, with the Glc-decorated hydrogels slightly stiffer at 37 °C and the Gal-decorated staying at a similar stiffness.

Cellular Studies with Peptides Decorated Before Hydrogel Formation

Acknowledgement: The cellular studies described in this subsection were carried out with Dr Bangfu Zhu (University of Cardiff).

The postulated role of carbohydrates in the development of oligodendrocytes (Sect. 1.4.4) made OPCs a suitable cell type to be cultured on the glycosylated hSAFs, formed with modifed hSAF-p1-A14Z. A hydrogel decorated with a combination of Gal and SGal should elicit a measurable cellular response. Pre-modified hSAF-p1-A14Z could readily be used to present multiple motifs, by mixing the different modified peptides in solution in the desired ratio before combination with hSAF-p2 to form the hydrogel. Assuming comparable rates of fibre formation for all hSAF-p1 derivatives, this should then give a statistical distribution of the two components, mimicking what would occur in the membrane of the myelin sheath with freely floating glycolipids.

Foregoing studies with RGD-modified hydrogels and PC12 cells show little difference in cellular response between 100% modification and 10% modification of the scaffold with the peptide [16, 19]. This stands to reason as there will be a maximum 'resolution' that cell surface receptors are able to interact with: for the hSAFs 10% decoration should result in motifs with an average spacing along the fibres of 40 nm, rather than a ≈4 nm separation for 100% decoration (a four heptad coiled coil is

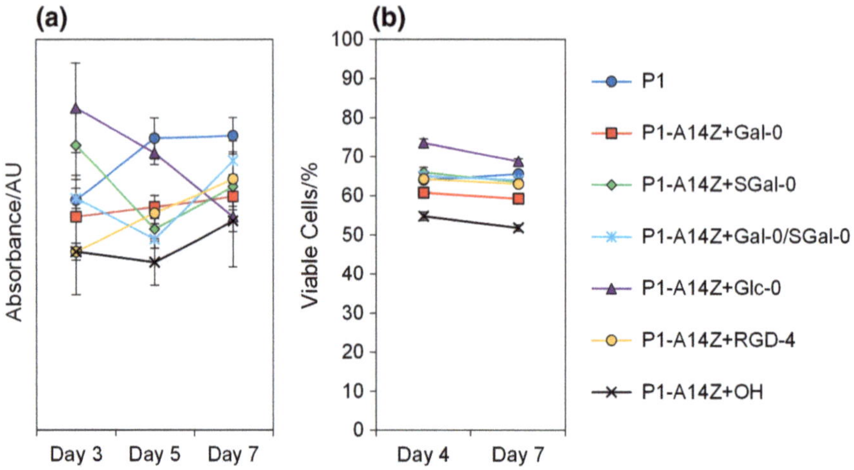

Fig. 5.11 Response of cultured murine oligodendrocytes to gels formed with hSAF-p2 and hSAF-p1 or hSAF-p1-K1Z decorated with simple carbohydrate modifiers or controls. **a** Viability assay; **b** Live/dead assay. Error bars are standard error

approximately 4 nm in length). Therefore, my studies with glycosylated hydrogels formed with hSAF-p1-A14Z used 10% total modified peptide, by pre-mixing in the correct ratio to hSAF-p1 in solution prior to combination with hSAF-p2 to form hydrogels.

Samples were made up with the Gal/SGal mixture, as well as Gal and SGal individually to test if either elicited a response alone. Glc-0-alkyne was also used as a separate control, as it has the same physical properties as Gal, but should not be bioactive due to its ubiquity as a metabolite. hSAF-p1-K1Z decorated with propargyl alcohol was used as a further negative control, as a peptide with the most simple motif that could be used to 'cap' the azide. This enabled me to ensure any response was not due to the triazole linker that is formed by CuAAC, or from the effects of the decoration procedure itself. A further set of gel samples were made decorated with an RGDS motif. hSAF-p1-A14Z decorated with the previously used RGD-peptide did not form hydrogels upon mixing with hSAF-p2, perhaps due to the steric bulk of the nonapeptide. However, when the five-amino-acid linker was replaced by tetraethylene glycol as a simple water-soluble linker with the same number of atoms as a spacer (to give RGD-4-alkyne), hydrogel formation was successful.[2]

Murine OPCs were cultured on hydrogels decorated with the various motifs and controls for seven days. Unfortunately, proliferation and viability assays showed no significant differences between any of the samples (Fig. 5.11). Total metabolic activity remained approximately the same across the seven days for all of the samples (Fig. 5.11a), as did the percentage of viable cells at around 60–70% (Fig. 5.11b). This could be due to the carbohydrate motifs of the propargyl glycosides not being

[2]See Sect. 5.6.1 for details of the synthesis of the linker used.

accessible to the cells, that is, they are buried within or too close to the peptide fibres. It is also possible that even if the motif did interact with the cells, this gave no cellular response. Previously the carbohydrates have been shown to affect differentiation rather than proliferation [3]. Attempts to characterise the differentiation of the cells cultured on the scaffolds proved unsuccessful.

5.5 Carbohydrate Modifiers with Varied Linkers

Acknowledgement: This work was carried out in conjunction with Ian Hazleden.

Although hSAF scaffolds could be modified with the propargyl glycosides, no cellular response to these new materials was observed for oligodendrocytes, even though a response had been anticipated. One possibility for this lack of response is that the close proximity of the appended motifs to the scaffold fibre might prevent binding to cell surface receptors. New carbohydrate modifiers were developed with longer, water-soluble linkers, in order to make the bioactive motifs more accessible. I also considered that the different linkers could be used to modulate the physical properties of the hydrogels—materials based on ethylene glycols are commonly used to make synthetic hydrogels themselves [20].

5.5.1 Synthesis of Carbohydrates with Extended Linkers

I chose linkers based on polyethers, as they are suitably water-soluble and should be biologically benign moieties [21]. I synthesised a range of linkers from ethylene (**34**), diethylene (**35**), and tetraethylene (**36**) glycols, with a terminal alkyne required for CuAAC at one end and free hydroxyl at the other so that they could be glycosylated analogously to the procedure used previously for propargyl alcohol. Using these species would give carbohydrate modifiers equivalent to the simple propargyl glycosides, but with additional separation from the peptide fibres ranging up to that given by the five-amino-acid linker used in the previous, successful, studies with bioactive peptides [16].

The linkers **37**, **38**, and **39** were made by alkylation of the ethylene glycols with propargyl bromide in water without requiring the previously reported anhydrous conditions [22] (Scheme 5.3). The ethylene glycols have two equivalent hydroxyls, so the desired product had to be separated from the dialkylated by-product.

Glycosylation of the synthesised linkers **38** and **39** with peracetylated Glc (**30**) and Gal (**31**) was carried out under the same conditions as for the synthesis of the simple carbohydrate modifiers with propargyl alcohol, with BF_3 acting as a Lewis acid (Scheme 5.4). Subsequent deprotection with catalytic NaOMe in MeOH afforded Glc-2-alkyne (**40**), Glc-4-alkyne (**41**), Gal-2-alkyne (**42**), and Gal-4-alkyne (**42**).

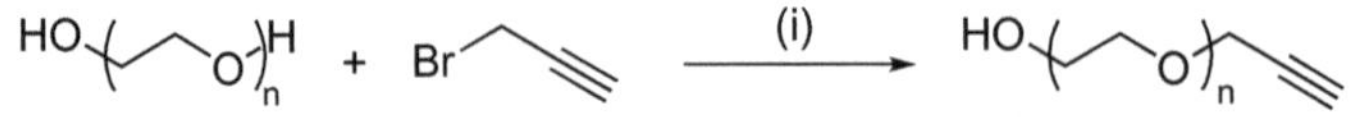

34 n = 1
35 n = 2
36 n = 4

37 n = 1 55%
38 n = 2 69%
39 n = 4 71%

Scheme 5.3 Synthesis of alkynyl polyether alcohols. Conditions: (i) 1.1 equiv. propargyl bromide, KOH (1.1 equiv.), H_2O, RT, 16 h

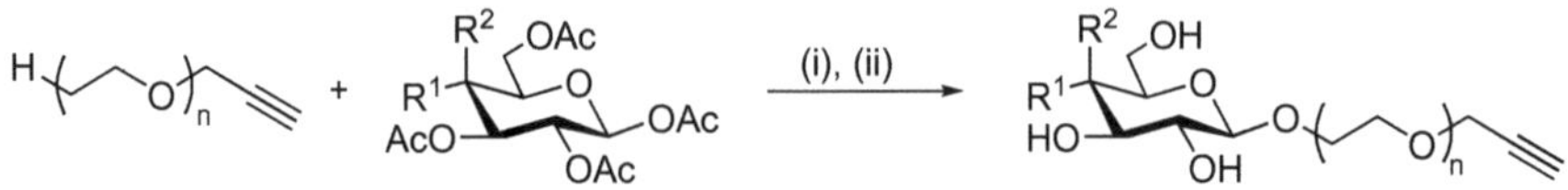

38 n = 2
39 n = 4

30 R^1 = OAc, R^2 = H
31 R^1 = H, R^2 = OAc

40 R^1 = OH, R^2 = H, n = 2 60%
41 R^1 = OH, R^2 = H, n = 4 60%
42 R^1 = H, R^2 = OH, n = 2 74%
43 R^1 = H, R^2 = OH, n = 4 79%

Scheme 5.4 Synthesis of Glc and Gal with alkynyl polyether linkers. Conditions: (i) 1.5 equiv. **38** or **39**, $BF_3.OEt_2$ (1.5 equiv.), anhydr. CH_2Cl_2, RT, 16 h. (ii) Cat. NaOMe, MeOH, RT, 10 min

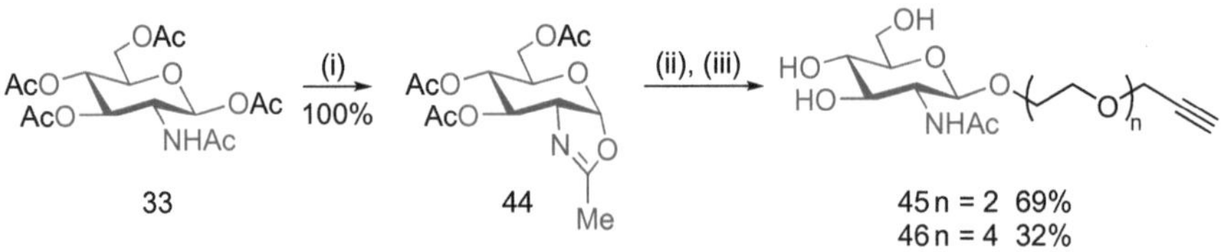

33

44

45 n = 2 69%
46 n = 4 32%

Scheme 5.5 Synthesis of GlcNAc with alkynyl polyether linkers. Conditions: (i) TMSOTf (1.1 equiv.), anhydr. CH_2Cl_2, reflux, 12 h. (ii) 2 equiv. **38** (n = 2) or **39** (n = 4), $CuCl_2$ (1.5 equiv.), anhydr. CH_2Cl_2, reflux, 24 h. (iii) Cat. NaOMe, MeOH, RT, 10 min

For GlcNAc, reaction of peracetylated glucosamine (**33**) with the ethylene glycol linkers was not successful under the conditions used previously for reaction with propargyl alcohol, with TMSOTf as a Lewis acid. Instead a by-product where the alcohol had reacted directly with TMSOTf was isolated, which may have been consuming the TMSOTf before activation of the carbohydrate was possible, and the alcohol was also being converted into a species that was non-nucleophilic. The TMSOTf forms an oxazoline intermediate (**44**), which is electrophilic at the anomeric position, to react with the alcohol. This oxazoline is quite stable, and so it was pre-formed by reaction with TMSOTf according to a published procedure [23] and isolated. It was then successfully reacted with the alkynyl linkers **38** and **39** using an alternative Lewis acid, $CuCl_2$, which did not react with the alcohol [24]. Deprotection with catalytic NaOMe in MeOH again gave GlcNAc-2-alkyne (**45**) and GlcNAc-4-alkyne (**46**).

By these methods, a small library of alkynyl monosaccharides with various ethylene glycol spacers was synthesised that is suitable for assessing the effect of linker length, both in terms of physical hydrogel properties and cellular response. For the latter, the Gal-based modifiers are probably the most useful, given the known response of hepatocytes to Gal [2], with the Glc derivatives providing an appropriate control.

5.6 A Library of Oligosaccharide Modifiers

Most of the signalling roles of carbohydrates in nature require the greater complexity of oligosaccharides over simple monosaccharides. Given their larger size, oligosaccharides may also have more substantial effects on the physical properties of hydrogels to which they are attached. Therefore, a series of oligosaccharides that are of biological relevance, along with appropriate controls, were synthesised by a combination of chemical and enzymatic synthesis to increase the utility of carbohydrate-modified scaffolds. First, monosaccharides were synthesised chemically that contained a terminal alkyne for CuAAC, a linker for spacing from the peptide fibres, and appropriate functionality to facilitate their use as glycosyl acceptors for carbohydrate-active enzymes. The monosaccharides were then built into the various oligosaccharides using enzymatic couplings.

Two classes of oligosaccharide were targeted. Given the role of Gal in liver and neural tissues (Sect. 1.4.4), the first class were two Gal-based oligosaccharides, α-1,4-D-galacatosyl D-galactose (diGal) (**47**) and Lac (**20**, Fig. 5.12).

The second class were β-galactosides LacNAc (**16**), Le^X (**17**), and SLe^X (**48**, Fig. 5.13), that bind to galectins, which are implicated in many roles including for neural cell types (Sect. 1.4.4).

Appropriate controls for use with both classes were also synthesised, in the form of Glc oligomers CelB (**18**) and cellotriose (CelT) (**49**, Fig. 5.14). These have very similar structures and so broadly the same chemical properties as the other neutral oligosaccharides, and so should have similar effects on the physical properties of the scaffolds, but are not known to cause any cellular response in the cell types being targeted.

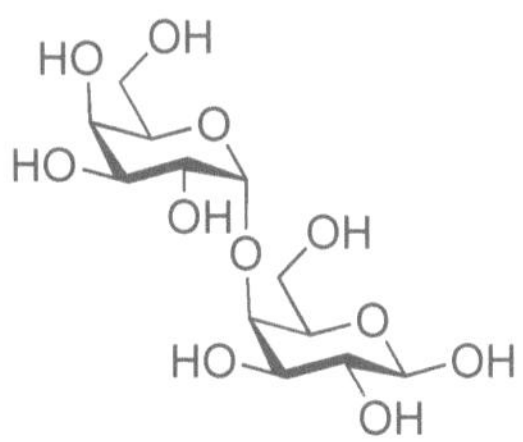
47

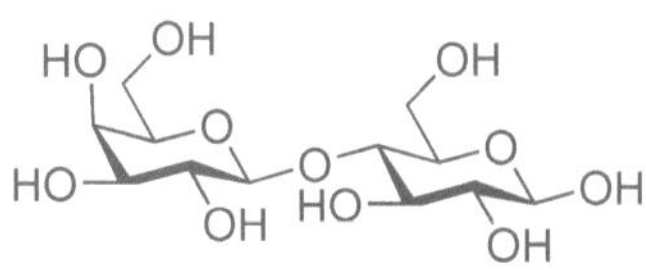
20

Fig. 5.12 Oligosaccharides diGal (**47**) and Lac (**20**)

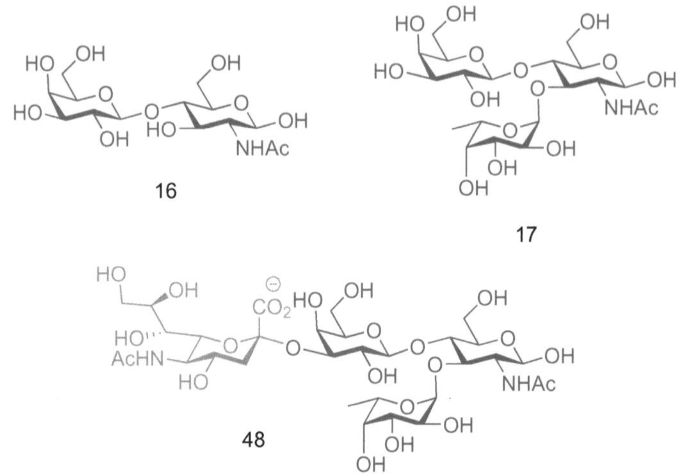

Fig. 5.13 β-Galactosides LacNAc (**16**), Le^X (**17**), and SLe^X (**48**)

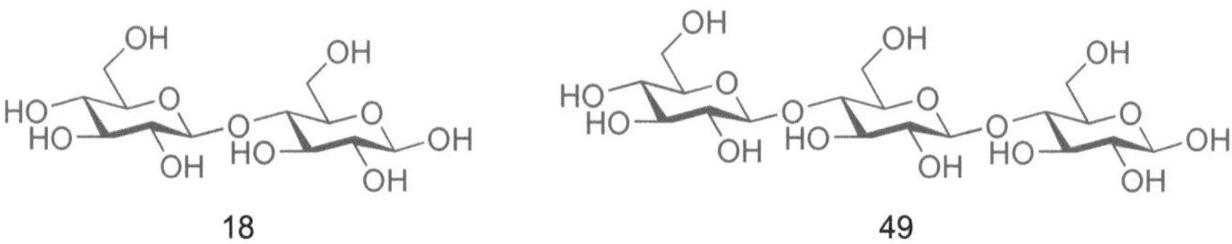

Fig. 5.14 Oligosaccharides CelB (**18**) and CelT (**49**)

5.6.1 Synthesis of Acceptors for Enzymatic Couplings

The enzymes of choice, glycosyltransferases and glycosynthases, are most tolerant to variation at the anomeric position of carbohydrates. Therefore, I installed the required spacer and alkynyl functionality at this site on my glycosyl acceptors. Moreover, glycosynthases show increased efficiency with acceptors that have aromatic groups at the anomeric position [25, 26], so I developed an adapted approach to the simple polyether linkers that included a benzene ring. These acceptors were used with all enzymatic couplings, to ensure any observed effects derived only from differences in the oligosaccharide moiety.

The approach chosen was adapted from a published synthesis of azide-functionalised monosaccharides [27]. Three acceptors were required to give the desired library of oligosaccharides, again Glc, Gal, and GlcNAc. *p*-Nitrophenyl glycosides are used in activity assays for glycosidases, meaning that several derivatives are available commercially, including the three monsaccharides of interest, and others can be synthesised from the peracetylated carbohydrates analogously to the propargyl glycosides. Therefore, these were used as starting points for the acceptor syntheses as they already had the aromatic moiety in place at the anomeric position.

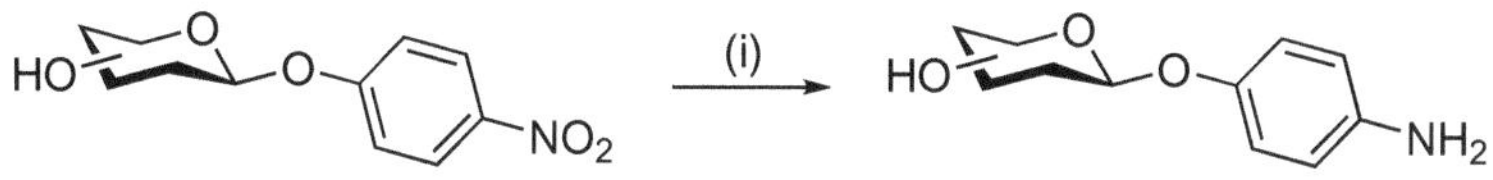

50 Glc
51 Gal
52 GlcNAc

53 Glc 100%
54 Gal 100%
55 GlcNAc 100%

Scheme 5.6 Synthesis of *p*-aminophenyl glycosides. Conditions: (i) H_2, cat. Pd/C, MeOH, RT, 16 h

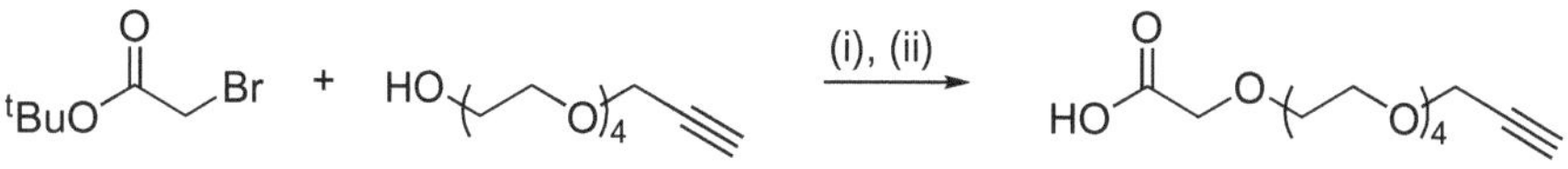

56

39

57 36%

Scheme 5.7 Synthesis of alkynyl polyether linker carboxylic acid. Conditions: (i) 2 equiv. *t*-butyl-α-bromoacetate (**56**), NaH (2 equiv.), anhydr. tetrahydrofuran(THF), RT, 16 h. (ii) TFA, CH_2Cl_2, RT, 1 h

p-Nitrophenyl glycosides **50**, **51**, and **52** were reduced to the corresponding *p*-aminophenyl glycosides **53**, **54**, and **55** under H_2 at atmospheric pressure with Pd on carbon as the catalyst [28] (Scheme 5.6), giving an amine handle available for further derivatisation.

The previously synthesised alkynyl polyether alcohol was made compatible with the *p*-aminophenyl glycosides by nucleophilic addition of the remaining free alcohol to *t*-butyl-α-bromoacetate (**56**), followed by hydrolysis of the ester with TFA (Scheme 5.7).

The amine-functionalised carbohydrates and acid-functionalised linkers were coupled through amide-bond formation, using the standard coupling conditions of *O*-benzotriazole-N,N,N',N'-tetramethyluronium hexafluorophosphate, (HBTU) and N,N-diisopropylethylamine (DIEA) in N,N-dimethylformamide (DMF). Amide linkages are more stable with respect to hydrolysis than other options, such as esters. Coupling was carried out with linkers derived from both diethylene glycol, which gives an overall linker that is the same length as the original pentapeptide linker or the simple tetraethylene glycol-derived linker alone, as well as from the tetraethylene glycol-derived linker itself to give **57**. The latter was used for subsequent studies in order to maximises spacing from the peptide. Forming the linkages with the unprotected *p*-aminophenyl glycosides **53**, **54**, and **55** was successful.

The final products were water soluble and were purified by reversed-phase HPLC, being easy to separate using gradients of MeCN in H_2O and detected by the UV absorbance of the aromatic moieties. As purifying unprotected carbohydrate derivatives is often difficult, this is a convenient property of the chosen linkers.

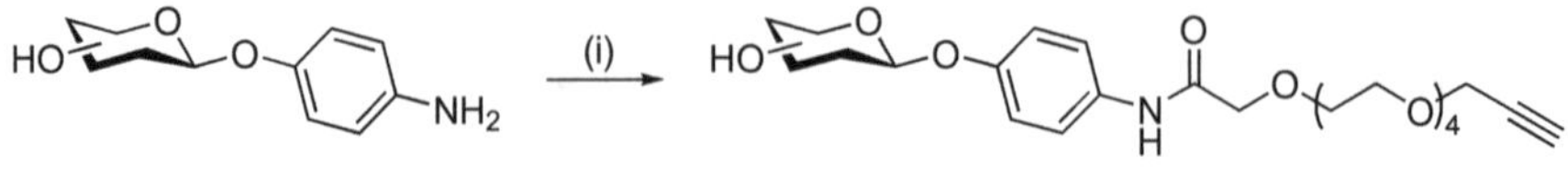

53 Glc	58 Glc 32%
54 Gal	59 Gal 39%
55 GlcNAc	60 GlcNAc 24%

Scheme 5.8 Synthesis of alkynyl glycosides with aromatic polyether linkers. Conditions: (i) 1.2 equiv. acid **57**, HBTU (1.18 equiv.), DIEA (2 equiv.), anhydr. DMF, RT, 16 h

5.6.2 Enzymatic Couplings

Acknowledgement: The work described in this subsection was carried out during a research placement in the lab of Prof. Stephen Withers (University of British Columbia, Vancouver, BC, Canada), and the enzymes and reagents were kindly donated by members of the Withers research group.

The synthesised glycosyl acceptors were then used to make a range of oligosaccharides using enzymatic couplings. Several enzymes were used, including glycosyltransferase wild-types and mutants designed to increase efficiency, and an engineered glycosynthase. The couplings were carried out in aqueous buffers determined by the enzyme being used, and were followed by thin layer chromatography (TLC). Products were confirmed by MS, purified by reversed-phase HPLC, and unambiguously identified by NMR spectroscopy. The latter was challenging, given the large number of C–H groups in similar environments (bound to one oxygen and either two carbons or a carbon and a proton), but was possible by identifying the characteristic shifts of the anomeric C–H protons.

In subsequent schemes the following abbreviation is used:

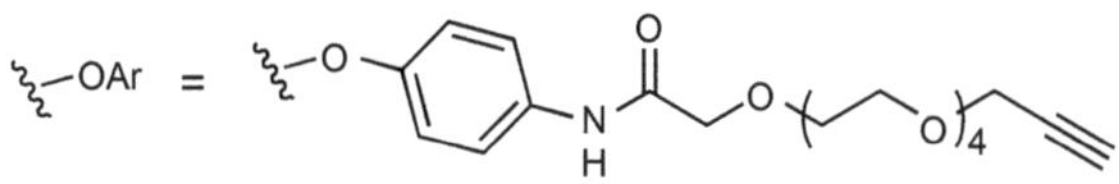

Synthesis of Cellobiose and Cellotriose

Functionalised CelB and CelT were synthesised as control oligosaccharides using a glycosynthase developed from a β-glucosidase from *Agrobacterium* sp., Abg 2F6 [26]. The wild-type enzyme is a retaining β-glycosidase, which has been converted into the bond-forming glycosynthase by removing a key nucleophilic Glu residue to give Abg-E358A [25]. Optimisation by random mutagenesis then gave the more efficient 2F6 form, which has four mutations in total from the wild type [26]. This enzyme promotes formation of a glycosidic linkage between an appropriate acceptor and a donor in the form of α-D-glucosy fluoride (GlcF) (**61**) [29]. This donor mimics the transition state of the glycosidase, with the anomeric C–F bond being broken by the acceptor hydroxyl. Reaction occurs exclusively at 4-hydroxyl position of glucose-type acceptors with this enzyme.

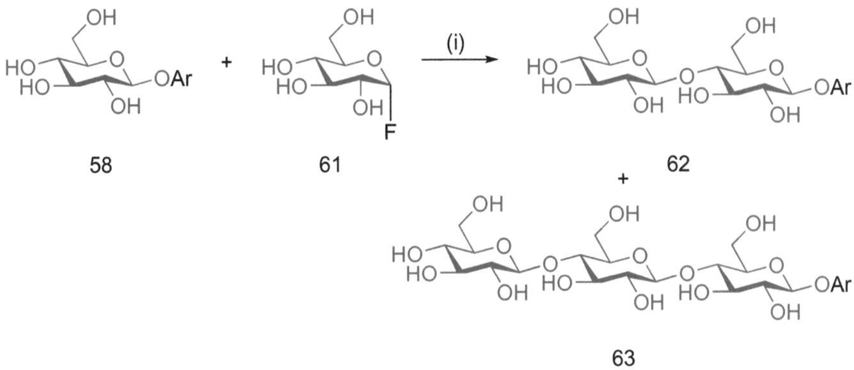

Scheme 5.9 Synthesis of CelB and CelT with alkynyl polyether linkers. Conditions: (i) 1.5 equiv. GlcF (61), Abg 2F6 (1 mg mL^{-1}), aq. NaPi buffer at pH 7.15, RT, 1 h

Reaction with the glucose acceptor **58** was achieved with the acceptor at a concentration of 10 nM and the donor **61** at 15 nM in sodium phosphate (NaPi) buffer at pH 7.15 (Scheme 5.9). It was carried out at 1 mL scale, so just over 5 mg of acceptor was used, and 1 mg of Abg 2F6 to give 1 mg mL^{-1} enzyme concentration. The reaction was followed by TLC, and after 1 h at room temperature (the optimal temperature for this enzyme) the starting material had been almost completely consumed. Precipitate was removed by centrifugation, and then the protein removed by passing over a C18 Sep-Pak. The product was eluted with a crude gradient of MeOH in H_2O, and observed by TLC in the 20 and 30% MeOH fractions. All subsequently synthesised oligosaccharides were purified using the same technique. This gave carbohydrates that were clean enough to identify by NMR spectroscopy, and to be used in further coupling steps. Reversed-phase HPLC was used to give the final, pure oligosaccharides for further studies.

More than one spot was by seen TLC after the reaction time, giving a 'ladder' of equally spaced spots, with the second and third from the top of the plate (the first being the starting material) being the most intense. This is because the product of the first coupling, **62**, also has a free 4-hydroxyl compatible with further enzymatic coupling, and the engineered enzyme binds a range of acceptors. Therefore, the product can react with a further donor molecule to give the trisaccharide **63**, corresponding to the third spot seen by TLC. Further reaction to give higher oligomers is in principle also possible, but given that only 1.5 equivalents of the donor was used the two major products were the di- and trisaccharides.

The synthesis of both oligosaccharides in one pot was not a problem, as both were desired to act as controls and the two were separable by HPLC. The final yield was 35% in terms of **62** and 10% in terms of **63**.

Synthesis of Lactose

Abg 2F6 is not only promiscuous in the acceptors it tolerates, but can also add Gal to acceptors by using α-D-galactosy fluoride (GalF) (**64**) as a glycosyl donor,

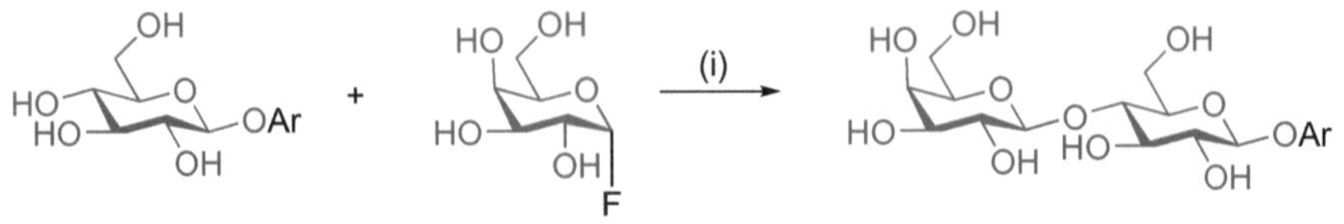

58 64 65

Scheme 5.10 Synthesis of Lac with alkynyl polyether linkers. Conditions: (i) 1.5 equiv. GalF (**64**), Abg 2F6 (1 mg mL^{-1}), aq. NaPi buffer at pH 7.15, RT, 1 h

although the reaction is less efficient [26]. The use of GalF with the glucoside **58** under otherwise identical conditions to the synthesis of CelB gave Lac (**65**) as a product in 52% yield (Scheme 5.10).

Unlike for CelB, the reaction product **65** does not have any free equatorial C4 hydroxyls, and so cannot act as an acceptor for further glycosylations. This means that Lac was the exclusive product in this case.

Synthesis of Digalactose

The disaccharide diGal was synthesised using the retaining galactosyltransferase LgtC, from *Neisseria meningitidis* [30]. This is a natural glycosyltransferase that uses a sugar nucleotide as the donor, in this case Gal with an anomeric uridine diphosphate (UDP) group, uridine-5′-diphosphogalactose (UDP-Gal) (**66**). As LgtC is a retaining transferase, and UDP-Gal occurs in the α-anomer, the enzyme catalyses the formation of α-glycosidic linkages, namely an α-1,4-galactosyl linkage onto the 4-hydroxyl of Gal. The catalysis is also metal-dependent, with a Mn^{2+} ion in the binding site [30].

The coupling was carried out using the conditions developed for kinetic studies of LgtC upon its first isolation [30]. The Gal-based acceptor **59** was used along with UDP-Gal as the donor (Scheme 5.11). 4-(2-hydroxyethyl)-1-piperazineethanesulfonic acid (HEPES) buffer was used at pH 7.5, and the reaction mixture incubated at 30 °C. $MnCl_2$ was included in the reaction mixture to ensure sufficient metal ions for the enzyme to be active. Alkaline phosphatase was also included, which is crucial to give a good yield of the product: UDP is a by-product of the reaction expelled by the donor upon formation of the glycosidic linkage, and the phosphatase breaks this down by cleaving the phosphate bonds [31], by Le Châtelier's principle driving the reaction forward to the product to replace the lost UDP. As with many enzymes, the presence of (BSA) reduces aggregation of the protein; enzymes are designed to work in the cellular environment, with many other proteins present.

As with all of the synthesised oligosaccharides, the diGal was isolated and purified using the same technique as CelB in a 48% yield. Although the product **67** had a free C4 hydroxyl, unlike with the CelB synthesis, only the disaccharide product was seen and no higher oligosaccharides. This is because LgtC, as the wild type, is less promiscuous than the Abg variant; it has evolved not to add additional Gal residues to digalactose motifs.

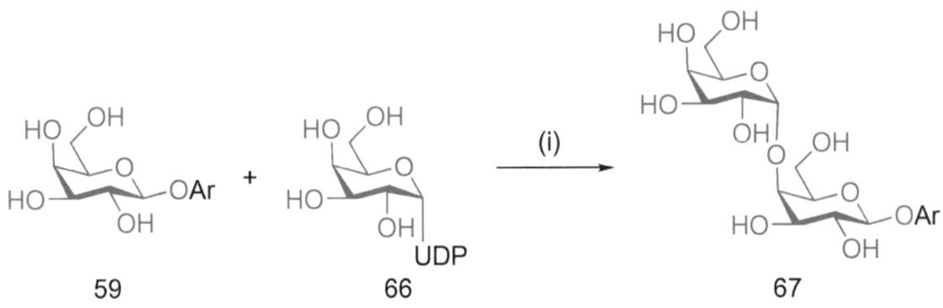

Scheme 5.11 Synthesis of diGal with alkynyl polyether linker. Conditions: (i) 1.5 equiv. UDP-Gal (**66**), LgtC, alkaline phosphatase, BSA, $MnCl_2$, KCl, dithiothreitol (DTT), aq. HEPES buffer at pH 7.5, 30 °C, 16 h

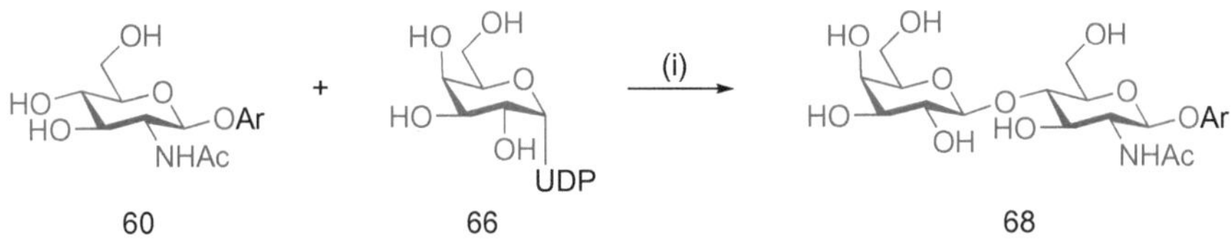

Scheme 5.12 Synthesis of LacNAc with alkynyl polyether linker. Conditions: (i) 1.2 equiv. UDP-Gal (**66**), GalT, alkaline phosphatase, BSA, $MnCl_2$, aq. Tris buffer at pH 7.4, 37°C, 16 h

Synthesis of *N*-Acetyllactosamine

β-1,4-Galactosyltransferase (GalT) is one of the most widely used glycosyltransferases, and is available commercially, and catalyses the addition of Gal onto the C4 hydroxyl of GlcNAc. I used GalT to make LacNAc, both as a sample for studies itself and for derivatisation with further carbohydrate couplings.

GlcNAc acceptor **60** was used, with UDP-Gal (**66**) as the donor (Scheme 5.12). GalT requires similar reaction conditions to LgtC, again being metal-dependent and so needing $MnCl_2$ in the buffer, as well as alkaline phosphatase and BSA. In this case, tris(hydroxymethyl)aminomethane(Tris) buffer was used at pH 7.4, and the enzyme was mammalian (isolated from bovine milk) meaning that the optimal reaction temperature is 37 °C. Purification by the techniques outlined previously gave the final product in a 72% yield.

An alternative method to synthesise LacNAc would be to use the enzyme Abg 2F6 with GalF as a donor and GlcNAc as an acceptor. Reaction was attempted with the GlcNAc acceptor and GalF donor, and evidence of reaction was seen by TLC and HPLC. However, NMR spectroscopy confirmed only approximately 50% conversion and so, given the success of the above transformation with GalT, this method was not pursued further.

Synthesis of Sialylgalactose

Sialic acids, especially NeuNAc, are commonly the terminal residues in naturally occurring glycans. I synthesised several sialylated oligosaccharides, working towards the synthesis of biologically relevant SLe^X [32]. This required the α-2,3-sialylation of

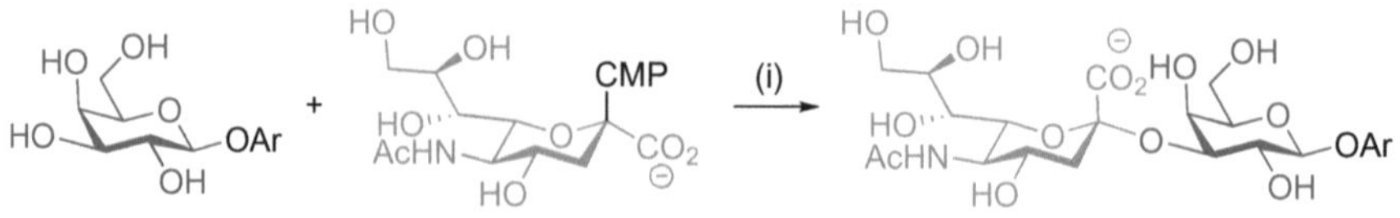

59 69 70

Scheme 5.13 Synthesis of SiaGal with alkynyl polyether linker. Conditions: (i) 1.5 equiv. NeuNAc-CMP (**69**), PmST1 M144D, $MgCl_2$, aq. Tris buffer at pH 8.5, 37 °C, 4 h

a Gal residue, which was achieved with the sialyltransferase PmST-1 from *Pasteurella multocida* [33]. This is an Mg-dependent glycosyltransferase, which catalyses the formation of both the 2,3-linkage and the 2,6-linkage. A recently developed mutant, M144D, favours formation of only the desired 2,3-linkage by altering the position of the glycosyl donor in the binding site [34]. The glycosyl donor for PmST1 is NeuNAc with (CMP) at the anomeric position, NeuNAc-CMP (**69**). This can be synthesised enzymatically prior to coupling or *in situ* in a multi-enzyme system [35].

Before carrying out sialylations on oligosaccharides that had been synthesised enzymatically, the enzyme and reaction conditions were verified by synthesising α-2,3-sialylgalactose (SiaGal) from the Gal-based acceptor **59**. As well as a proof-of-concept, this product is also of use as a control, as alone it should not be biologically active but it contains the negatively charged sialic acid group, which may have effects on the material properties that can be assessed independent of biological activity. The coupling was carried out in Tris buffer at pH 8.5 and 37 °C (Scheme 5.13). The yield for this coupling was 35%.

Synthesis of Sialyl-*N*-acetyllactosamine

There are two possible approaches to the synthesis of SLe^X. The first is to sialylate Le^X, which is possible with PmST1 M144D [34]. The second is to fucosylate sialyl-*N*-acetyllactosamine (SiaLacNAc), which is also possible with the appropriate fucosyltransferase [36, 37]. The second approach was chosen as the reagents and enzyme for the sialylation were in hand and the FucT still required expression. Although SiaLacNAc does not have any known desirable biological function, it would also prove a useful control for effects on physical propertis.

SiaLacNAc was synthesised by carrying out sialylation of the LacNAc-based acceptor **68** by the same method used for making SiaGal (Scheme 5.14). This again used the enzyme PmST1 M144D, with NeuNAc-CMP (**69**) as the donor. This reaction was more successful than for the coupling to the single Gal, with a final yield of 58%.

α-1,3-Fucosyltransferase Expression

While all of the enzymes described thus far were kindly provided by members of the Withers research group, the fucosyltransferase needed to make Le^x and SLe^x was not available. The required enzyme, an α-1,3-fucosyltransferase (FucT) from *Helicobacter pylori* was therefore expressed and purified. A construct optimised to

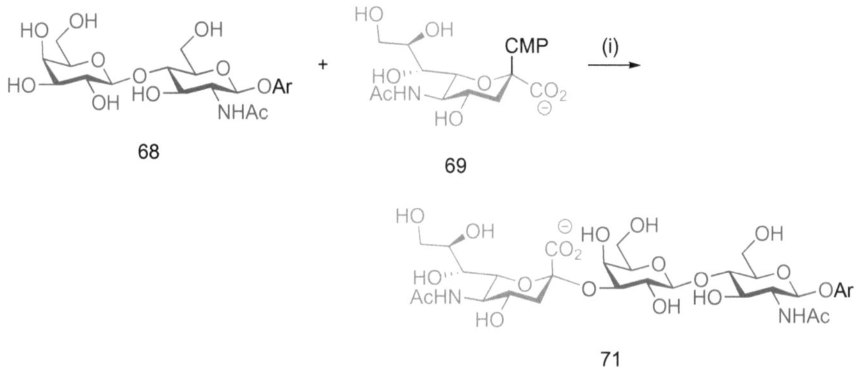

Scheme 5.14 Synthesis of α-2,3-sialyl-*N*-acetyllactosamine (SiaLacNAc) with alkynyl polyether linker. Conditions: (i) 1.5 equiv. NeuNAc-CMP (**69**), PmST1 M144D, $MgCl_2$, aq. Tris buffer at pH 8.5, 37 °C, 4h

increase solubility [37] was used; this optimisation has been achieved by removing a hydrophobic tail and tandem repeat section, with activity maintained.

The plasmid containing the construct was transformed into an ampicillin-resistant strain of *E. coli*, and a starter culture grown overnight in ampicillin-containing LB medium at 37 °C. The next day, 500 mL of medium was inoculated with the starter culture and grown for 2 h before the temperature was reduced to 25 °C and expression induced by the addition of IPTG. After a further 5 h, the cells were collected by centrifugation and lysed. The cell lysate was purified by FPLC using a Ni-affinity column, as the FucT construct was His-tagged. The purification was successful, giving about 6 mg of the desired enzyme from the 500 mL culture.

Synthesis of Lewis^X

Le^X has been identified as a possible marker of neural stem cells [38], making it of interest for neural tissue engineering. Le^X was synthesised from the LacNAc derivative (**68**) using the expressed FucT, which transfers a fucose residue onto the 3-hydroxyl of a GlcNAc through an α-linkage (Scheme 5.15) [37]. The natural donor for fucosyltransferases is fucose with a guanosine diphoshate (GDP) group at the anomeric position, (GDP-Fuc) (**72**). The product (**73**) was obtained after reaction at 25°C. FucT is another example of a Mn-dependent glycosyltransferase, so $MnCl_2$ was used in the reaction buffer, and again addition of alkaline phosphatase increases the yield of the product by degrading the GDP by-product. The yield for this step was 43%.

Synthesis of Sialyl Lewis^X

The FucT from *Helicobacter pylori* can also be used to fucosylate SiaLacNAc to give SLe^X [37]. This was carried out using conditions identical to those for the fucosylation of LacNAc (Scheme 5.16) to give the tetrasaccharide **74**. This final step

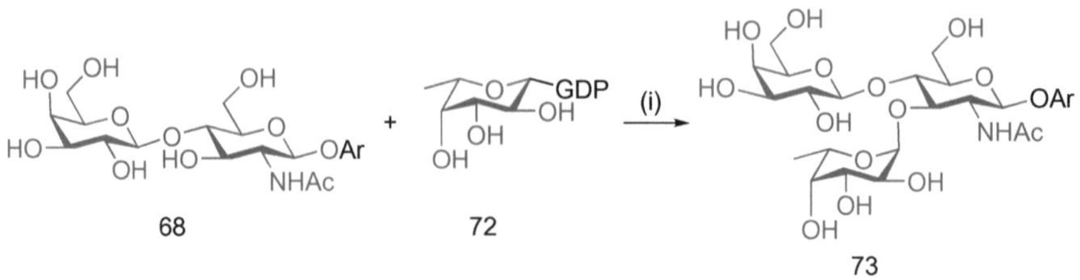

Scheme 5.15 Synthesis of Le^X with alkynyl polyether linker. Conditions: (i) 1.2 equiv. GDP-Fuc (**72**), FucT, alkaline phosphatase, BSA, $MnCl_2$, aq. Tris buffer at pH 7.4, 25 °C, 2 h

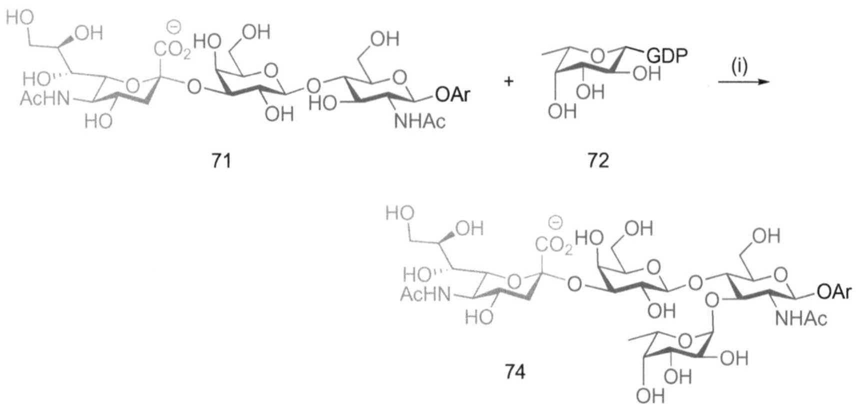

Scheme 5.16 Synthesis of SLe^X with alkynyl polyether linker. Conditions: (i) 1.2 equiv. GDP-Fuc (**72**), FucT, alkaline phosphatase, BSA, $MnCl_2$, aq. Tris buffer at pH 7.4, 25 °C, 2 h

in the tetrasaccharide synthesis was achieved in a very similar yield to that for the coupling to give Le^X, at 41%.

The synthesis of this tetrasaccharide in three steps from the chemically synthesised monosaccharide acceptor highlights the utility of enzymatic couplings for building complex oligosaccharides. Although development, expression, and purification of the required enzymes, and synthesis of the donor carbohydrates, is by no means trivial, it still represents a viable alternative to chemical synthesis.

I have used these enzymatic techniques to obtain a library of carbohydrates suitable for application to the engineering of several tissue types. A combination of wild-type and mutated glycosyltransferases and a glycosynthase have been used to synthesise the oligosaccharides, functionalised with alkynes so that they are compatible with the azide-containing hSAFs. This enables the investigation of the properties of hydrogels decorated with the bioactive carbohydrates, as well as the response of the targeted cell types.

References

1. Werz DB, Ranzinger R, Herget S, Adibekian A, von der Lieth C-W, Seeberger PH (2007) ACS Chem Biol 2:685–691
2. Cho CS, Seo SJ, Park IK, Kim SH, Kim TH, Hoshiba T, Harada I, Akaike T (2006) Biomaterials 27:576–585
3. Boggs JM, Gao W, Zhao J, Park H-J, Liu Y, Basu A (2010) FEBS Lett 584:1771–1778
4. Kornfeld R, Kornfeld S (1985) Annu Rev Biochem 54:631–664
5. Varki A, Cummings RD, Esko JD, Freeze HH, Stanley P, Bertozzi CR, Hart GW, Etzler ME (eds) (2009) Essentials of glycobiology, 2nd edn. Cold Spring Harbor Laboratory Press, Cold Spring Harbor (NY)
6. Toole BP (2004) Nat Rev Cancer 4:528–539
7. Mereyala HB, Gurrala SR (1998) Carbohydr Res 307:351–354
8. Vauzeilles B, Dausse B, Palmier S, Beau J-M (2001) Tetrahedron Lett 42:7567–7570
9. van Kasteren SI, Kramer HB, Jensen HH, Campbell SJ, Kirkpatrick J, Oldham NJ, Anthony DC, Davis BG (2007) Nature 446:1105–1109
10. Zhao J, Liu Y, Park H-J, Boggs J M, Basu A, Program F (2012) Bioconjug Chem 1–64
11. Guo J, Ye X-S (2010) Molecules 15:7235–7265
12. Izumi M, Fukase K, Kusumoto S (2002) Biosci Biotechnol Biochem 66:211–214
13. Merkel L, Beckmann HSG, Wittmann V, Budisa N (2008) ChemBioChem 9:1220–1224
14. Mosmann T (1983) J Immunol Methods 65:55–63
15. Russo L, Gloria A, Russo T, D'Amora U, Taraballi F, De Santis R, Ambrosio L, Nicotra F, Cipolla L (2013) RSC Adv 3:6286–6289
16. Mehrban N, Abelardo ES, Wasmuth A, Hudson KL, Mullen LM, Thomson AR, Birchall MA, Woolfson DN (2014) Adv Healthc Mater 3:1387–1391
17. Mehrban N, Zhu B, Tamagnini F, Young FI, Wasmuth A, Hudson KL, Thomson AR, Birchall MA, Randall AD, Song B, Woolfson DN (2015) ACS Biomater Sci Eng 1:431–439
18. Mahmoud ZN, Gunnoo SB, Thomson AR, Fletcher JM, Woolfson DN (2011) Biomaterials 32:3712–3720
19. Kantlehner M, Schaffner P, Finsinger D, Meyer J, Jonczyk A, Diefenbach B, Nies B, Hölzemann G, Goodman SL, Kessler H (2000) ChemBioChem 1:107–114
20. Slaughter BV, Khurshid SS, Fisher OZ, Khademhosseini A, Peppas NA (2009) Adv Mater 21:3307–3329
21. Niidome T, Yamagata M, Okamoto Y, Akiyama Y, Takahashi H, Kawano T, Katayama Y, Niidome Y (2006) J Control Release 114:343–347
22. Kislukhin AA, Higginson CJ, Hong VP, Finn MG (2012) J Am Chem Soc 134:6491–6497
23. Nakabayashi S, Warren CD, Jeanloz RW (1986) Carbohydr Res 150:C7–C10
24. Wittmann V, Lennartz D (2002) Eur J Org Chem 1363–1367
25. Mackenzie LF, Wang Q, Warren RAJ, Withers SG (1998) J Am Chem Soc 7863:5583–5584
26. Kim Y-W, Lee SS, Warren RAJ, Withers SG (2004) J Biol Chem 279:42787–42793
27. Fukuda T, Onogi S, Miura Y (2009) Thin Solid Films 518:880–888
28. Goebel WF, Avery OT (1929) J Exp Med 50:521–531
29. Williams SJ, Withers SG (2000) Carbohydr Res 327:27–46
30. Persson K, Ly HD, Dieckelmann M, Wakarchuk WW, Withers SG, Strynadka NC (2001) Nat Struct Biol 8:166–175
31. Unverzagt C, Kunz H, Paulson J (1990) J Am Chem Soc 01:9308–9309
32. Phillips ML, Nudelman E, Gaeta FC, Perez M, Singhal AK, Hakomori S-I, Paulson JC (1990) Science 250:1130–1132
33. Yu H, Chokhawala H, Karpel R, Yu H, Wu B, Zhang J, Zhang Y, Jia Q, Chen X (2005) J Am Chem Soc 127:17618–17619
34. Sugiarto G, Lau K, Qu J, Li Y, Lim S, Mu S, Ames JB, Fishe AJ, Chen X (2012) ACS Chem Biol 7:1232–1240
35. Rich JR, Cunningham A-M, Gilbert M, Withers SG (2011) Chem Commun (Cambridge, U.K) 47:10806–10808

36. Dumas DP, Ichikawa Y, Wong C-H, Lowe JB, Nair RP (1991) Bioorg Med Chem Lett 1:425–428
37. Lin S-W, Yuan T-M, Li J-R, Lin C-H (2006) Biochemistry 45:8108–8116
38. Hennen E, Faissner A (2012) Int J Biochem Cell Biol 44:830–833

Part III
Conclusions and Supplementary Information

Chapter 6
Conclusions and Future Directions

6.1 Protein–Carbohydrate Interactions

Conclusions

To summarise the first part of this thesis, I have demonstrated the utility of protein X-ray crystal structures containing carbohydrates as a source of information on PCIs. I generated a database of coordinates of carbohydrate residues and proximal amino acids from structures in the PDB [1], with the carbohydrates identified using GlyVicinity, [2] and the quality of the data validated using Privateer [3]. Analysis of the distribution of amino acids in the carbohydrate binding sites, and particularly comparison of the distributions around different monosaccharides, are powerful tools for understanding carbohydrate-based interactions at the molecular level.

The analysis shows that hydrophobic amino acids are generally disfavoured in carbohydrate-binding sites, except for those with aromatic side chains, which are over-represented. This is in accordance with the recognised phenomenon of carbohydrate–aromatic interactions [4]. Carbohydrate C–H bonds interact with amino-acid functional groups differently between and within monosaccharides. Some show a marked preference for aromatic residues, and this is at the expense of interactions with non-aromatic hydrophobics.

With this confirmed importance of carbohydrate–aromatic interactions, I investigated the electronic properties of the carbohydrates and aromatics to explain the differences in interactions for different species. Computational ESP calculations for the carbohydrates show that the C–H protons have different electrostatic potentials within and between monosaccharides. I parametrised and identified all of the CH–π interactions involving the monosaccharide residues in the data set. Quantifying the participation of the different carbohydrate C–H bonds in CH–π interactions shows that the most-positive C–H protons are involved in CH–π interactions the most frequently, supporting a key role for the electrostatic component of carbohydrate–aromatic interactions.

K. L. Hudson, *Carbohydrate-Based Interactions at the Molecular and the Cellular Level*, Springer Theses, https://doi.org/10.1007/978-3-319-77706-1_6

The participation of aromatic amino acids in CH–π interactions also supports this finding. The most electron-rich aromatic system, that of Trp, was the most common CH–π acceptor, and Try was favoured over Phe.

The importance of electrostatics in carbohydrate–aromatic interactions was confirmed by NMR spectroscopic studies of solution-phase associations. Again, the extent of interaction with aromatic species of different C–H bonds correlates with the electronics of the C–H bond, as well as the propensities in the crystallographic data set. Interaction is also dependent on the electronics of the aromatic system, with no interaction observed for indoles with a neutral electrostatic potential over the ring.

Significance

These results are of importance for drug design and other areas, as CH–π interactions are implicated in many aspects of chemistry and biology [5]. The interactions between carbohydrates and proteins play many biological roles, including several related to health (Sect. 1.2.1) [6], and there is increasing interest in so-called 'glycomimetic' drugs designed to inhibit or co-opt carbohydrate-binding proteins. Understanding what drives carbohydrate–aromatic interactions, which are known to be key to carbohydrate binding, should guide the design of better drugs. The many PCIs that have been shown to involve CH–π interactions [4] can be targeted by designing glycomimetics with more electropositive C–H protons. These would interact with aromatic amino acids, particularly Trp, more favourably than the native carbohydrates.

Appreciation of the importance of electrostatics in carbohydrate–aromatic interactions may also lead to further insights into binding modes of lectins, and also the mechanisms of carbohydrate-active enzymes. Factors that enhance the positivity of C–H bonds could be crucial to energetically favourable carbohydrate binding by proteins through enhanced CH–π interactions. For example, coordination of hydroxyls to a divalent metal cation will withdraw electron density from the carbohydrate and so further polarise C–H bonds. Also, enzymes enhance reaction rates by stabilising transition states, and the mechanisms of many carbohydrate-active enzymes involve the build-up of positive charge on the carbohydrate ring [7]. This will increase the electropositivity of the transition state as a whole, and so enhance electrostatics in any CH–π interactions. This could be a way in which carbohydrate–aromatic interactions influence enzyme activity—indeed, recently the polarity of carbohydrate C–H bonds has been shown to be key to the function of a glycosylase involved in DNA repair [8]. The deoxyribose C–H bonds of methylated DNA nucleotides are more electropositive than for non-methylated nucleotides, and this increased polarity is recognised through CH-π interactions with Trp residues to trigger excision and repair.

The implications of these findings extend beyond PCIs. Carbohydrates serve as a useful model for ligand binding due to their common occurrence in the PDB and predictable conformation, but the principles about aromatic interactions should be applicable to all ligands. Understanding the nature of CH–π interactions also has broader significance as they are important in many other areas, from protein structure to reaction mechanisms [5].

Future Directions

Applying what was learned about the nature of carbohydrate–aromatic interactions to the design of glycomimetics and determining the effects on binding to lectins would be a useful extension of this work. Carbohydrate analogues with more-positive C–H bonds can be accessed through the synthesis of species with electron-withdrawing groups such as halides. The converse should also be possible, in that more-electron-rich aromatic groups should enhance binding, for example in synthetic lectins. Interactions should also be weakened by the opposite trends. One possible set of experiments to probe this would be the mutagenesis of a key aromatic residue in the carbohydrate-binding site of a lectin for unnatural amino acids with electron-deficient aromatic rings to quantify changes in binding constant. A similar approach has been taken to establish the importance of cation–π interactions in nicotinic acetylcholine receptors [9].

More in-depth studies are also possible with the database of CH–π interactions. By separating monosaccharide residues and ignoring substitution and position in oligosaccharide in this study, a lot of detail was lost for the sake of sample size. Also, I only looked at a subset of possible monosaccharide residues, ignoring those with charged substituents such as uronic acids and amino sugars as well as non-pyranose forms. Extending the study to these classes could provide further insights. An analysis to determine the influence of participation of carbohydrate hydroxyls in hydrogen bonding or coordination to a cation on the formation of CH–π interactions may expand the utility of the data.

It is also possible to investigate further classes of interaction using the data set, for example hydrogen bonding itself. This would allow a general picture of all aspects of carbohydrate-based interactions to be elucidated, which could lead to the determination of idealised binding sites for specific carbohydrates. These could then be mutated into protein scaffolds to give *de novo* lectins, or identified patterns used to scan for unknown binding sites in existing protein structures.

6.2 Carbohydrates in Tissue Engineering

Conclusions

The second project presented herein demonstrates that the hSAF system is a useful scaffold for the development of tissue engineering technologies. The parent hydrogels have favourable properties for mammalian cell culture [10], and a modifiable azide-containing variant allows for functionalisation with biological cues to improve performance [11, 12]. This study expanded the utility of the modifable version to provide a system for the assessment of carbohydrates as functional motifs in tissue engineering.

I synthesised several model alkyne-functionalised monosaccharides, and with these verified the glycosylation of azide-functionalised hSAFs. The ligation of carbohydrate moieties by CuAAC is successful and does not alter the key properties of the hydrogels, proving their usefulness. Preliminary studies of fibroblasts cultured on carbohydrate-modified scaffolds show that mammalian cells remain viable on the system, and even suggest a possible beneficial effect on cellular development through increased viability and proliferation.

Moving the azide-containing residue to the middle of one of the peptides allows linkage of carbohydrates prior to hydrogel formation. This simplifies the formation of functionalised scaffolds, especially with combinations of biological cues. This system was used to investigate the effect of targeted monosaccharides on the development of OPCs, although no cellular response was seen in studies completed to this point. This could be due to the short linkers between the carbohydrates and the scaffolds preventing the interactions with the cells required to induce a response.

To address this I synthesised a library of glycosides with water-soluble linkers of various lengths and terminal alkynes suitable for CuAAC, which can be used to study the effect of linker length and type on cellular response, and also may be useful for modulating the scaffold properties. I also used enzymatic couplings to access several suitably functionalised oligosaccharides that are applicable to tissue engineering and can be used for other studies of carbohydrate-based interactions. The library includes carbohydrates that should give a response in neural and hepatic cell lines, as well as appropriate controls for effects on physical properties.

Significance

The foremost contribution of these results is to demonstrate the use of the hSAF system to apply carbohydrates to tissue engineering. The many biological roles that carbohydrates are involved in makes them potentially very useful in this field. As hSAFs are a flexible and modular system, appropriate carbohydrates can be combined with other factors, for example appendage of functional peptide motifs [11] and tuned physical properties [13]. The development of a system that can be functionalised prior to hydrogel formation should also facilitate the practical application of hSAFs. This research shows that carbohydrate-modified scaffolds are compatible with mammalian cell culture, and indeed may lead to non-specific favourable responses at the least.

The library of alkyne-functionalised carbohydrates that I synthesised can be used to study effects on cells, both generally and specifically, either with hSAFs or other compatible scaffolds. Indeed, the modifiers are compatible with any azide-functionalised system for which appendage of carbohydrates could effect a response. Enzymatic coupling has been shown to be an accessible and robust technique for the straightforward synthesis of a variety of functional oligosaccharides.

Future Directions

The next stage with this project is to carry out further cell culture experiments with the carbohydrates and the targeted cell types. The most likely effect of the carbohydrates targeting oligodendrocytes is to influence differentiation [14], so successful experiments to assess this for cells cultured on the modified materials would be useful. Given the known response of liver cells to Gal-functionalised materials [15], the Gal-based monosaccharides provide useful tools for determining the optimal linker length by analysing cultured hepatocytes. The effects of several of the synthesised oligosaccharides on cultured neural cells would also be of interest, to investigate if the targeting of galectins induces favourable responses in terms of proliferation or differentiation, for example.

It would also be worthwhile to develop methods of functionalisation other than CuAAC. Strain-promoted (copper-free) azide-alkyne cycloaddition would be preferable, as it would be quicker and not involve toxic and damaging copper species [16]. Photochemical techniques, such as thiol-ene 'click' chemistry, could also be useful [17]. This could be achieved with only natural amino acids (Cys), and allow spatial patterning of functional groups to direct tissue development. The developed alkyne-functionalised carbohydrates would also be compatible with this technique, as thiol-yne coupling is also possible [18].

A more general long-term benefit of this project will hopefully be the realisation of the power of using carbohydrates to enhance tissue engineering scaffolds.

References

1. Berman HM, Westbrook J, Feng Z, Gilliland G, Bhat TN, Weissig H, Shindyalov IN, Bourne PE (2000) Nucl Acids Res 28:235–242
2. Lütteke T, Frank M, von der Lieth C-W (2005) Nucl Acids Res 33:D242–D246
3. Agirre J, Iglesias-Fernández J, Rovira C, Davies GJ, Wilson KS, Cowtan KD (2015) Nat Struct Mol Biol 22:833–834
4. Asensio JL, Ardá A, Cañada FJ, Jiménez-Barbero J (2013) Acc Chem Res 46:946–954
5. Nishio M (2011) Phys Chem Chem Phys 13:13873–13900
6. Walt D, Aoki-Kinoshita KF, Bendiak B, Bertozzi CR, Boons G-J, Darvill A, Hart G, Kiessling LL, Lowe J, Moon RJ, Paulson JC, Sasisekharan R, Varki AP, Wong C-H (2012) Transforming glycoscience: a roadmap for the Future. National Academies Press, Washington, D.C
7. Vasella A, Davies GJ, Böhm M (2002) Curr Opin Chem Biol 6:619–629
8. Mullins EA, Shi R, Parsons ZD, Yuen PK, David SS, Igarashi Y, Eichman BF (2015) Nature 527:254–258
9. Beene DL, Brandt GS, Zhong W, Zacharias NM, Lester HA, Dougherty DA (2002) Biochemistry 41:10262–10269
10. Banwell EF, Abelardo ES, Adams DJ, Birchall MA, Corrigan A, Donald AM, Kirkland M, Serpell LC, Butler MF, Woolfson DN (2009) Nat Mater 8:596–600
11. Mehrban N, Abelardo ES, Wasmuth A, Hudson KL, Mullen LM, Thomson AR, Birchall MA, Woolfson DN (2014) Adv Healthc Mater 3:1387–1391
12. Mehrban N, Zhu B, Tamagnini F, Young FI, Wasmuth A, Hudson KL, Thomson AR, Birchall MA, Randall AD, Song B, Woolfson DN (2015) ACS Biomater Sci Eng 1:431–439
13. Wasmuth A, Mehrban N, Hudson KL, Thomson AR, Mullen LM, Birchall MA, Woolfson DN, Manuscript in preparation
14. Boggs JM, Gao W, Zhao J, Park H-J, Liu Y, Basu A (2010) FEBS Lett 584:1771–1778
15. Cho CS, Seo SJ, Park IK, Kim SH, Kim TH, Hoshiba T, Harada I, Akaike T (2006) Biomaterials 27:576–585
16. van Hest JCM, van Delft FL (2011) ChemBioChem 12:1309–1312
17. Dondoni A (2008) Angew Chem Int Ed 47:8995–8997
18. Minozzi M, Monesi A, Nanni D, Spagnolo P, Marchetti N, Massi A (2011) J Org Chem 76:450–459

Chapter 7
Experimental Details and Supplementary Data

7.1 The Protein–Carbohydrate Interaction Database

7.1.1 Generation of Data Set

To identify carbohydrate residues contained within protein X-ray crystal structures from the PDB, an output was generated from GlyVicinity [1, 2] (accessed 6th Novemeber 2013). The parameters were set to identify all carbohydrates from ligands and glycans in all X-ray crystal structures with resolution ≤2.5 Å along with all amino acids with at any atom within 4 Å of any carbohydrate atom. The output was saved and processed in HTML format.

For the initial studies, redundancy of protein structures in the data set was addressed using PISCES [3]. A list was created of all of the PDB crystal structures identified in the GlyVicinity output as containing carbohydrates. These were submitted to the PISCES server [4] at maximum mutual sequence identity of 40%, culled by chain, with all other parameters set as the defaults. Only carbohydrates from PDB entries on the culled list were then considered. For the final study, separate lists were created for the PDB entries containing each of the monosaccharide residues of interest. These lists were then each submitted to a redundancy cull using CD-HIT [5, 6], with a maximum mutual sequence identity of 95%. Only residues from the first PDB entry of each cluster of similar structures in the CD-HIT output was used for subsequent analysis.

Separate working files containing the coordinates, in PDB format, of the carbohydrate and proximal amino acids found by GlyVicinity were generated for all monosaccharides in the GlyVicinity output for which this was possible. The GlyVicinity data was separated by PDB entry and then by oligosaccharide chains within that entry. The carbohydrate and amino acid coordinates for each monosaccharide in each of these was generated as a separate file by cross-referencing the residue labels from GlyVicinity with the coordinates contained within coordinate files obtained from the PDB. For cases where the residue label for the identified monosaccharide did not

K. L. Hudson, *Carbohydrate-Based Interactions at the Molecular and the Cellular Level*, Springer Theses, https://doi.org/10.1007/978-3-319-77706-1_7

correspond to the carbohydrate, it was identified by the atom label where possible, otherwise that example was discarded. Monosaccharides from cases where whole oligosaccharides were identified by a single residue label in the crystal structure were also discarded. Multiple examples of the same interaction from oligomeric crystal structures were prevented by comparing the identity and residue labels of the proximal amino acids for each monosaccharide within a PDB entry; cases where these matched a previously processed example were discarded.

Acknowledgement: The validation of carbohydrate strucutres in the PDB was completed in collaboration with Dr Jon Agirre (University of York).

To validate the quality of data used for interpretation, an analysis was carried out across the PDB for residues with the labels identified as the monosaccharides of interest (by GlyVicinity) using the programme Privateer [7]. A list of validated residues was then generated using the following parameters: RSCC $\geq$0.8, crystal structure resolution $\leq$2 Å, conformation $^1C_4$ or $^4C_1$. Only residues on this validated list were used for subsequent analysis.

Investigation of specific monosaccharide classes was achieved by identifying the appropriate coordinate files using the GlyVicinity assignment. Monosaccharide residues were determined as being from ligands or glycans using the chain LINUCS assigned to the overall oligosaccharide by GlyVicinity.

Files in PDB format, containing the coordinates from protein X-ray crystal structures, were visualised using the programme PyMOL [8]. Multiple examples of monosaccharides were visualised by alignment of the coordinates on the carbohydrate ring atoms.

7.1.2 Interpretation of Data Set

Composition of Amino Acid Side Chains

The composition of amino acids proximal to monosaccharide residues was derived from the composition of amino acid in the database of minimal PDB files for the monosaccharide(s) in question, i.e., determined by the GlyVicinity output after accounting for errors, redundancy, and duplicates.

Distribution of Amino Acid Side Chains

To determine the population of sectors around carbohydrates by different amino acids, the side chains of the proximal amino acids in the working coordinate files were treated as single points, defined as the mid-point of selected atoms (Table 7.1) for each amino acid. The sector was then assigned for each mid-point by finding the closest carbohydrate atom, of the carbons or the ring oxygen. The position relative to a plane through C1, C3, and C5 of the carbohydrate was also recorded. The mid-points were also used for the visualisation of side-chain distribution, displayed as spheres.

Table 7.1 Atoms, by PDB atom label, used to determine amino acid side chain 'centres'

Amino acid	Atoms
Ala	CB
Arg	NE, CZ, NH1, NH2
Asn	CG, OD1, ND2
Asp	CG, OD1, OD2
Cys	SG
Gln	CD, OE1, NE2
Glu	CD, OE1, OE2
Gly	CA
His	CG, ND1, NE2, CD2, CE1
Ile	CG1, CG2, CD1
Leu	CG, CD1, CD2
Lys	NZ
Met	SD, CE
Phe	CG, CD1, CD2, CE1, CE2, CZ
Pro	CB, CG, CD
Ser	OG
Thr	OG1, CG2
Val	CB, CG1, CG2
Trp	CG, CD1, CD2, NE1, CE2, CE3, CZ2, CZ3, CH2
Tyr	CG, CD1, CD2, CE1, CE2, CZ, OH

Identification of CH–π Interactions

Parameters were used to identify CH–π interactions as defined in Fig. 2.7a. These were adapted from those used in a published study of CH–π interactions in protein crystal structures [9]. Parameters were calculated relative to carbohydrate C–H carbons and amino acid aromatic rings. For Trp, the five- and six-membered rings were treated separately, as TrpA and TrpB, respectively. The CH–π distance was defined as the distance between the centre of the aromatic ring and the C–H bearing carbon, and was required to be $\leq$4.5 Å. The CH–π angle was defined as the angle between the plane of the aromatic ring and the plane through C1, C3, and C5 of the carbohydrate for axial C–H bonds, and the carbohydrate plane $+70.5°$ for equatorial C–H bonds, and was required to be $\leq 40°$. This parameter was not used for exocyclic C–H carbons. The C-projection distance was defined as the distance between the projection of the C–H carbon onto the plane of the aromatic ring and the aromatic ring's centre, and was required to be $\leq$1.6 Å for His and TrpA and $\leq$2.0 Å for Phe, TrpB, and Tyr. When multiple C–H carbons satisfied these criteria for the same aromatic ring, that with the shortest C-projection distance was selected as having the interacting C–H bond.

To aid visualisation, files in PDB format were generated containing the coordinates each monosaccharide and any aromatic amino acids engaged in CH–π interactions.

7.2 Crystallographic Database Analysis Supplementary Data

7.2.1 Composition of Amino Acids Around Glycans

Monosaccharide	L-Fuc	D-GlcNAc	D-Man		All
Anomer	α	β	α	β	
Number	12	438	64	15	529
Proximal Amino Acids					
Total	39	1812	224	33	2108
% Aromatic	10	15	15	21	15
% Aliphatic	18	22	26	27	23
% Polar	64	61	57	51	62
Average	3.25	4.14	3.50	2.20	3.98
Standard Deviation	1.86	1.94	1.48	1.47	
α-Face Amino Acids					
Average	0.5	1.5	0.9	0.8	1.4
% Aromatic	17	10	05	25	10
% Aliphatic	17	09	17	17	10
% Polar	67	81	78	58	80
β-Face Amino Acids					
Average	0.6	0.7	0.7	0.2	0.7
% Aromatic	0	13	26	0	14
% Aliphatic	43	28	22	0	27
% Polar	57	56	52	100	56
CH–π interactions					
Total	0	40	2	2	44
Average	0	0.09	0.03	0.13	0.08

7.2.2 Distribution of Amino Acids Around Monosaccharides

Monosaccharide structures with C–H protons labelled systematically; spatial distribution of centres of side-chains, represented as spheres, around all examples of each monosaccharide from the data set overlaid; proportion of amino acid side chains nearest to each carbohydrate carbon that are aromatic and aliphatic on each face. Adapted from Hudson et al. [10].

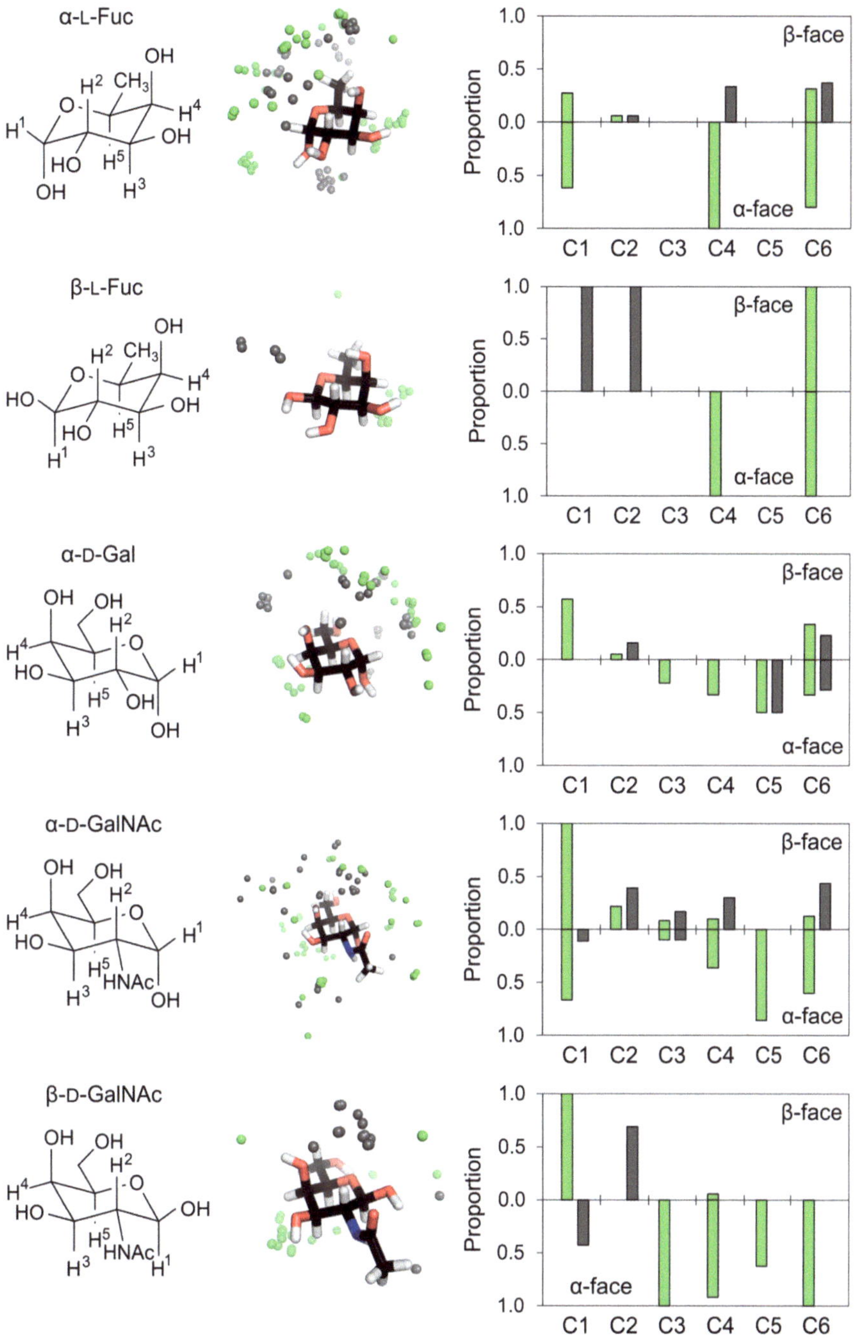

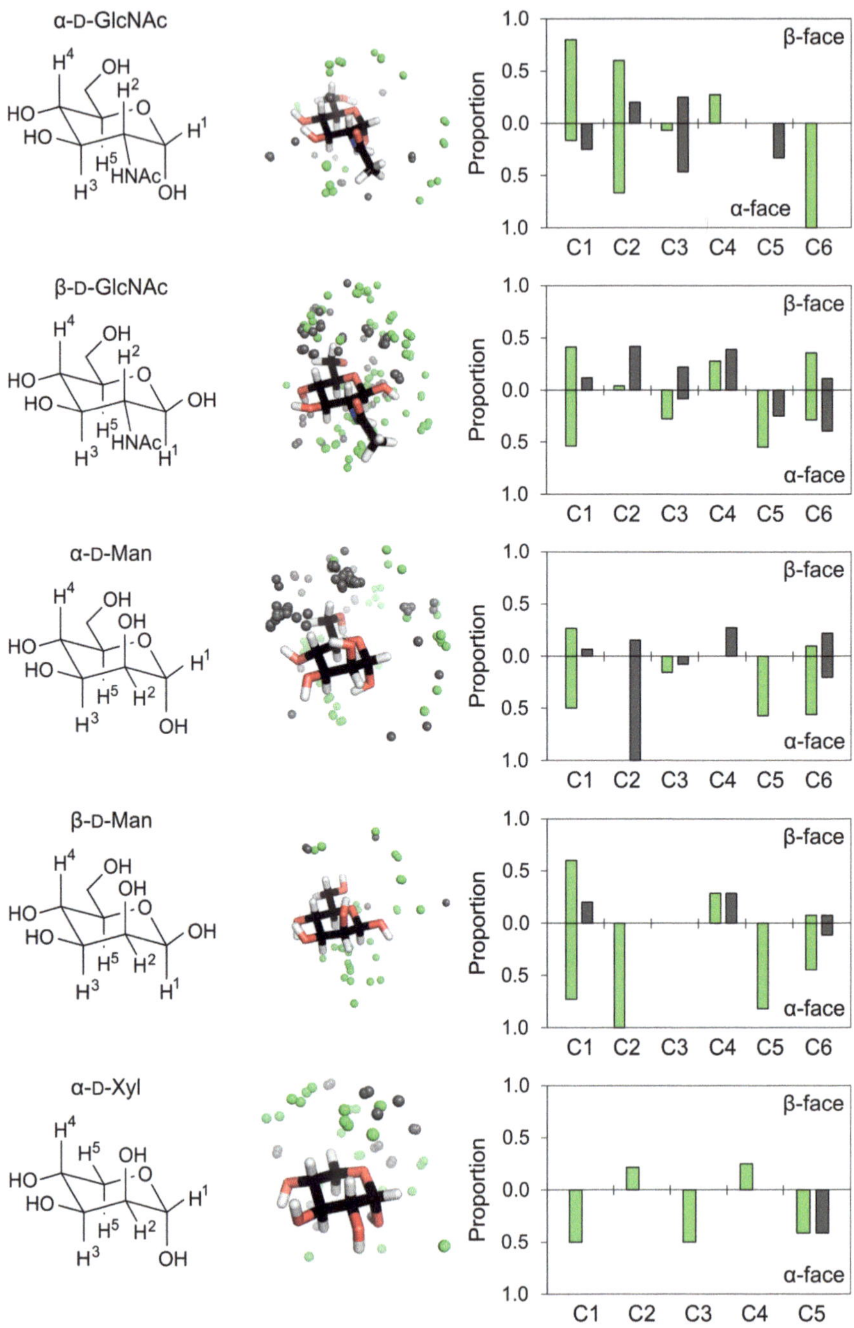

α-D-GlcNAc
β-D-GlcNAc
α-D-Man
β-D-Man
α-D-Xyl
HO
OH
HNAc
O
Proportion
1.0
0.5
0.0
0.5
1.0
β-face
α-face
C1
C2
C3
C4
C5
C6

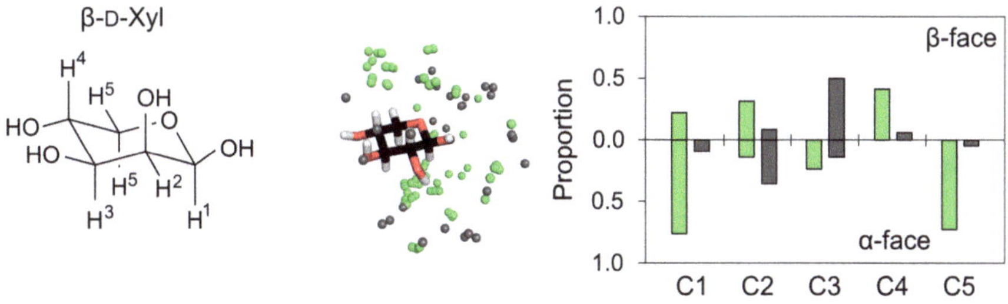

7.2.3 *CH–π Interactions of Monosaccharide Residues*

(**a**) Three views of optimised structure. (**b**) Corresponding views of ESP calculated for optimised structure. (**c**) Corresponding views of all monosaccharide residues in data set overlaid, with aromatic rings of amino acids identified as participating in CH–π interactions shown in green. (**d**) Average involvement of C–H protons in CH–π interactions. (**e**) Composition of proximal aromatic amino acids (white pie chart), with proportion involved in CH–π interactions shaded as green wedges (by area) (Figs. 7.1, 7.2, 7.3, 7.4, 7.5, 7.6, 7.7, 7.8 and 7.9).

7.3 NMR Studies of Carbohydrate–Aromatic Interactions

All chemicals used were purchased directly (except for β-methyl-D-mannopyranoside), used as purchased and of at least 97% purity. Solutions were prepared on a w/v basis. ^{1}H NMR spectra were acquired in D_2O on a Bruker Avance-500 500 MHz spectrometer with a DCH cryoprobe. Experiments used a spectral window from 11 to −1 ppm, a 4 s acquisition time, a 2 s relaxation delay, and 64 scans. NMR experiments with a relaxation delay of 15 s were run to verify indole concentration. The shift of the trimethyl peak of DSS was normalized to $\delta_{DSS} = 0$ ppm. For the data points shown, three series of experiments were conducted at the same glycoside and indole concentrations: indole only, glycoside only, and mixed samples. The chemical shifts were averaged over three replicates, and the chemical-shift perturbations were reported as $\Delta\delta = \delta_{indole} - \delta_{indole\text{-}free}$.

Methyl-β-D-mannopyranoside was synthesised by Dr Robert Brown (University of Wisconsin-Madison, Madison, WI, USA) by catalytic hydrogenation of methyl-3,4,6-tri-*O*-benzyl-β-D-mannopyranoside [10], which was synthesized according to literature procedures [11, 12].

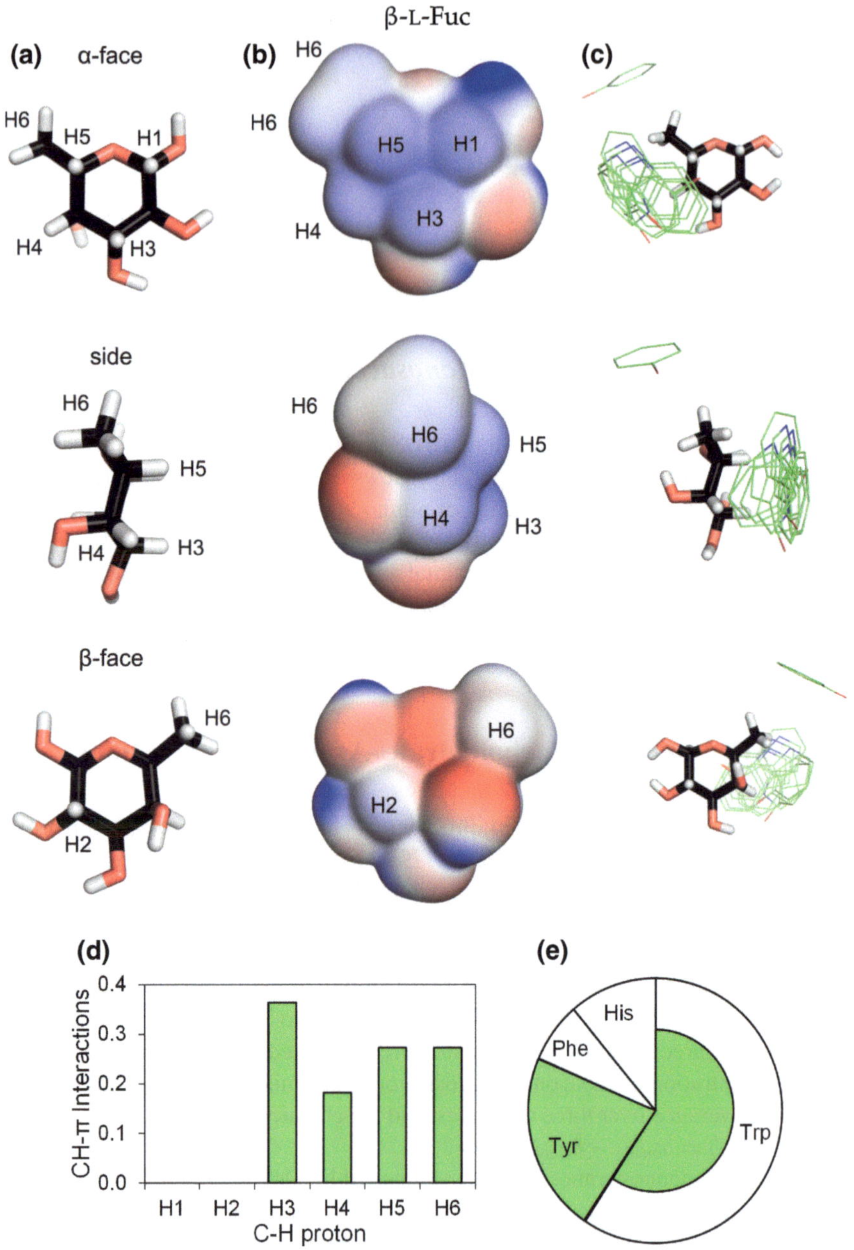

Fig. 7.1 CH?π interactions of β-L-Fuc in data set. **a** Three views of optimised structure. **b** ESP calculated for optimised structure. **c** All monosaccharide residues in data set overlaid, with identified amino acids CH?π acceptor aromatic rings shown in green. **d** Average involvement of C?H protons in CH?π interactions. **e** White pie chart: proximal aromatic amino acids; Green wedges: those involved in CH?π interactions

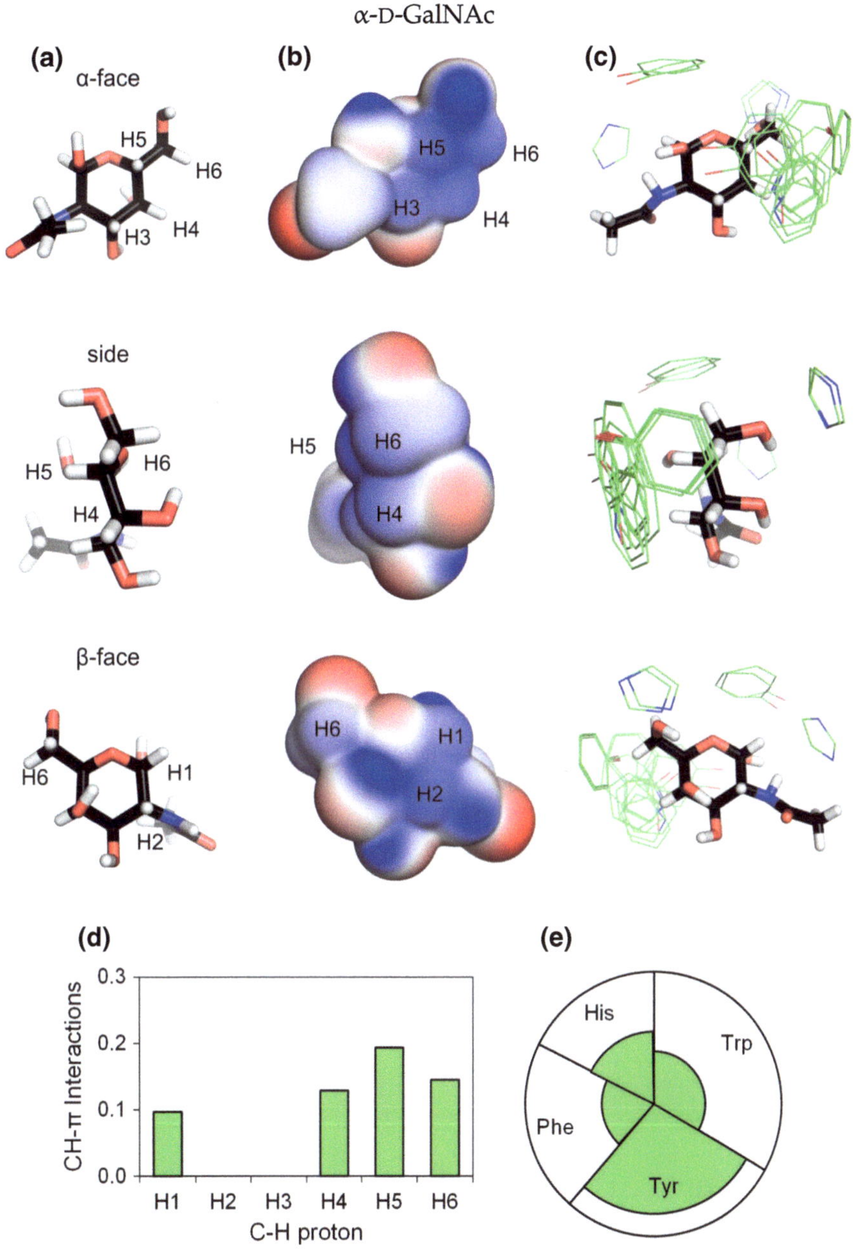

Fig. 7.2 CH?π interactions of α-D-GalNAc in data set. **a** Three views of optimised structure. **b** ESP calculated for optimised structure. **c** All monosaccharide residues in data set overlaid, with identified amino acids CH?π acceptor aromatic rings shown in green. **d** Average involvement of C?H protons in CH?π interactions. **e** White pie chart: proximal aromatic amino acids; Green wedges: those involved in CH?π interactions

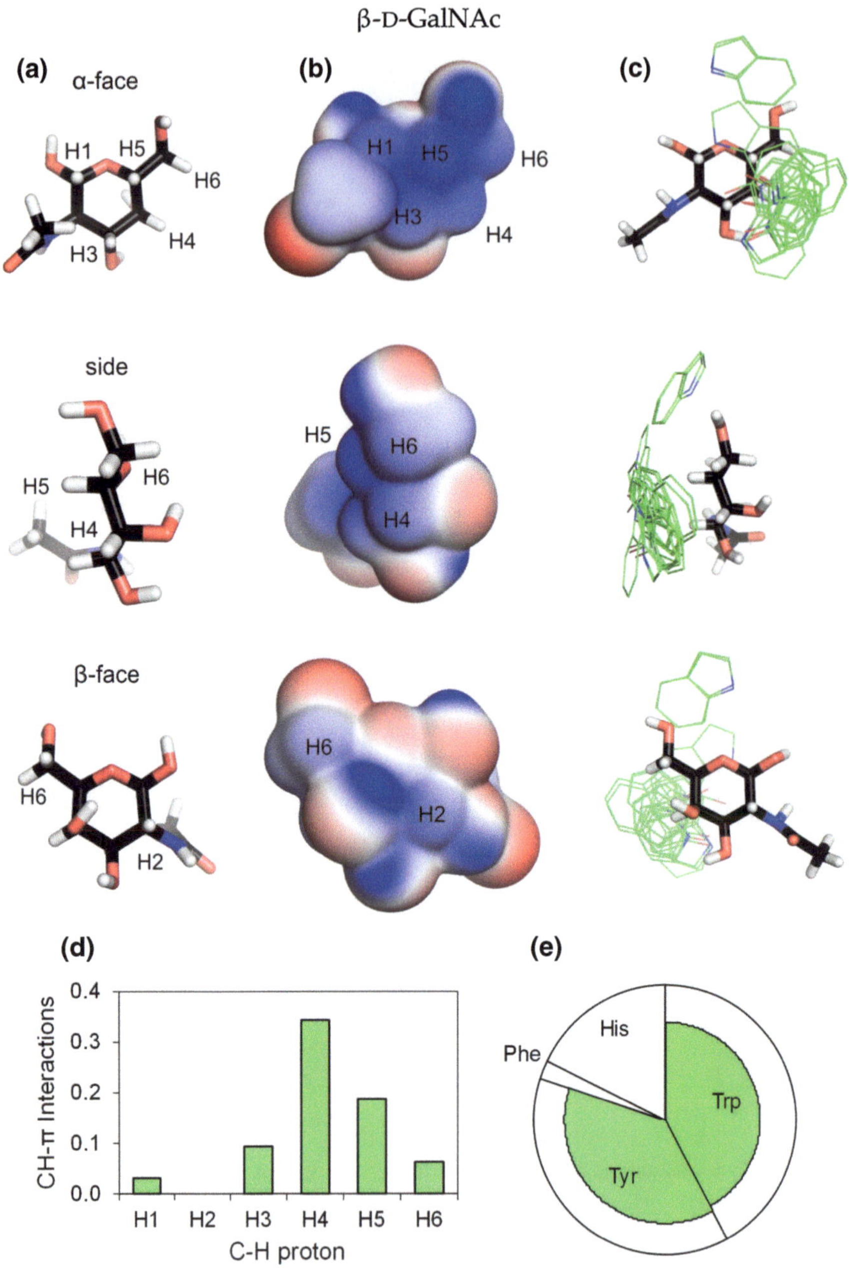

Fig. 7.3 CH?π interactions of β-D-GalNAc in data set. **a** Three views of optimised structure. **b** ESP calculated for optimised structure. **c** All monosaccharide residues in data set overlaid, with identified amino acids CH?π acceptor aromatic rings shown in green. **d** Average involvement of C?H protons in CH?π interactions. **e** White pie chart: proximal aromatic amino acids; Green wedges: those involved in CH?π interactions

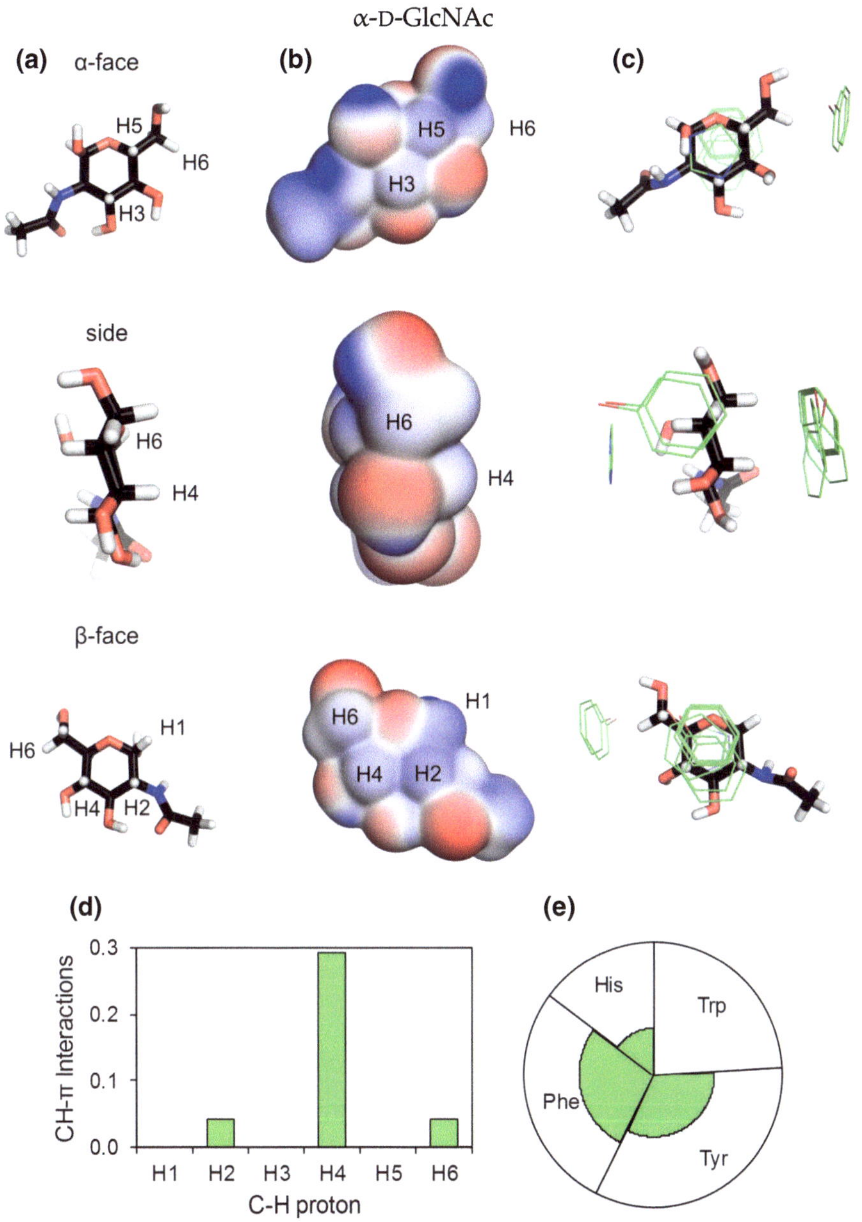

Fig. 7.4 CH?π interactions of α-D-GlcNAc in data set. **a** Three views of optimised structure. **b** ESP calculated for optimised structure. **c** All monosaccharide residues in data set overlaid, with identified amino acids CH?π acceptor aromatic rings shown in green. **d** Average involvement of C?H protons in CH?π interactions. **e** White pie chart: proximal aromatic amino acids; Green wedges: those involved in CH?π interactions

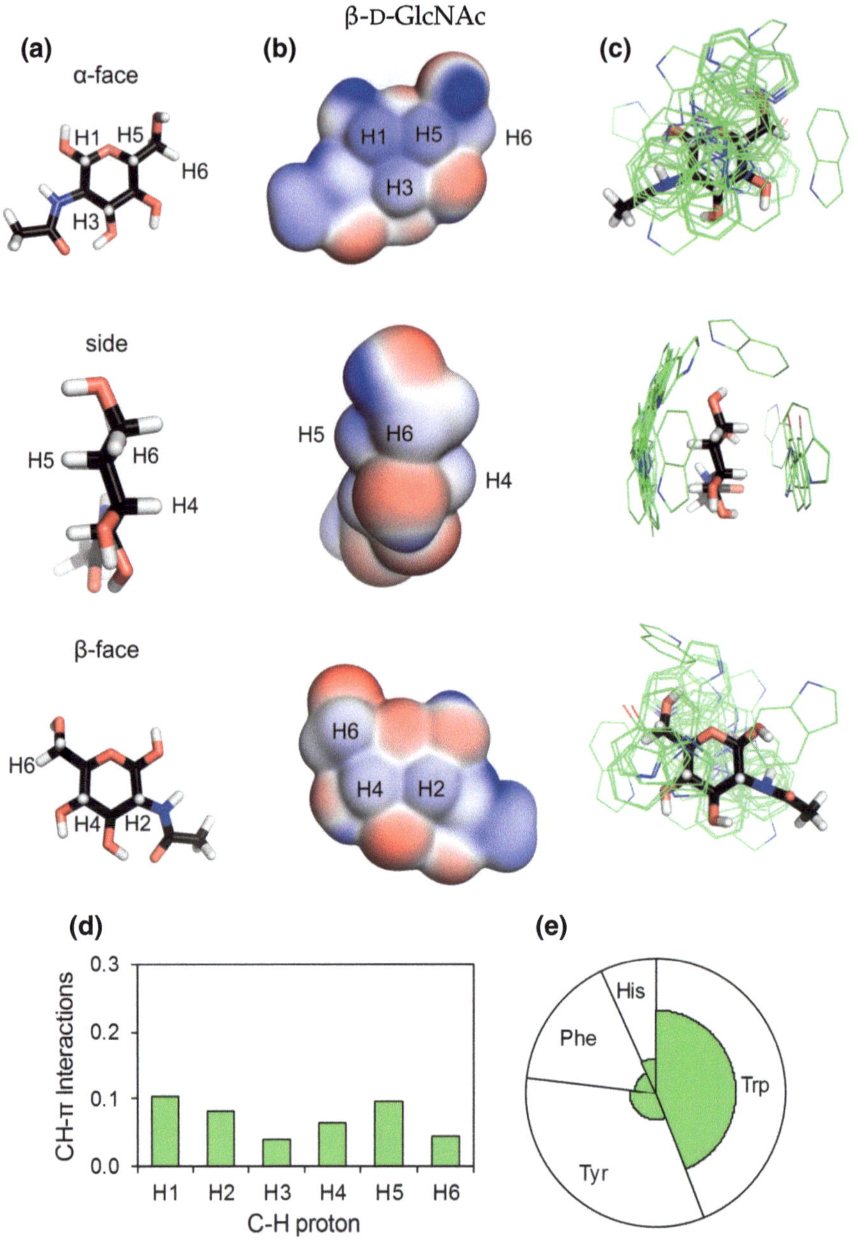

Fig. 7.5 CH?π interactions of β-D-GlcNAc in data set. **a** Three views of optimised structure. **b** ESP calculated for optimised structure. **c** All monosaccharide residues in data set overlaid, with identified amino acids CH?π acceptor aromatic rings shown in green. **d** Average involvement of C?H protons in CH?π interactions. **e** White pie chart: proximal aromatic amino acids; Green wedges: those involved in CH?π interactions

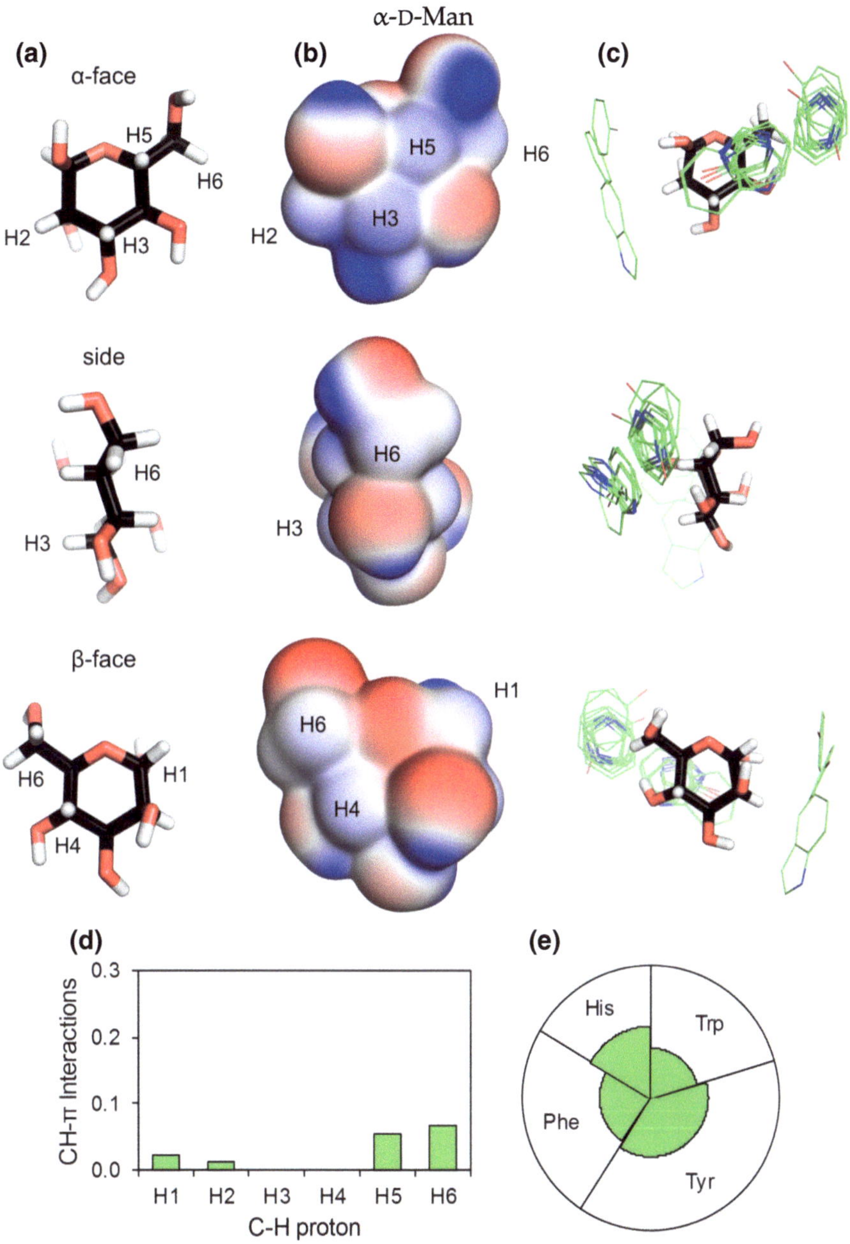

Fig. 7.6 CH?π interactions of α-D-Man in data set. **a** Three views of optimised structure. **b** ESP calculated for optimised structure. **c** All monosaccharide residues in data set overlaid, with identified amino acids CH?π acceptor aromatic rings shown in green. **d** Average involvement of C?H protons in CH?π interactions. **e** White pie chart: proximal aromatic amino acids; Green wedges: those involved in CH?π interactions

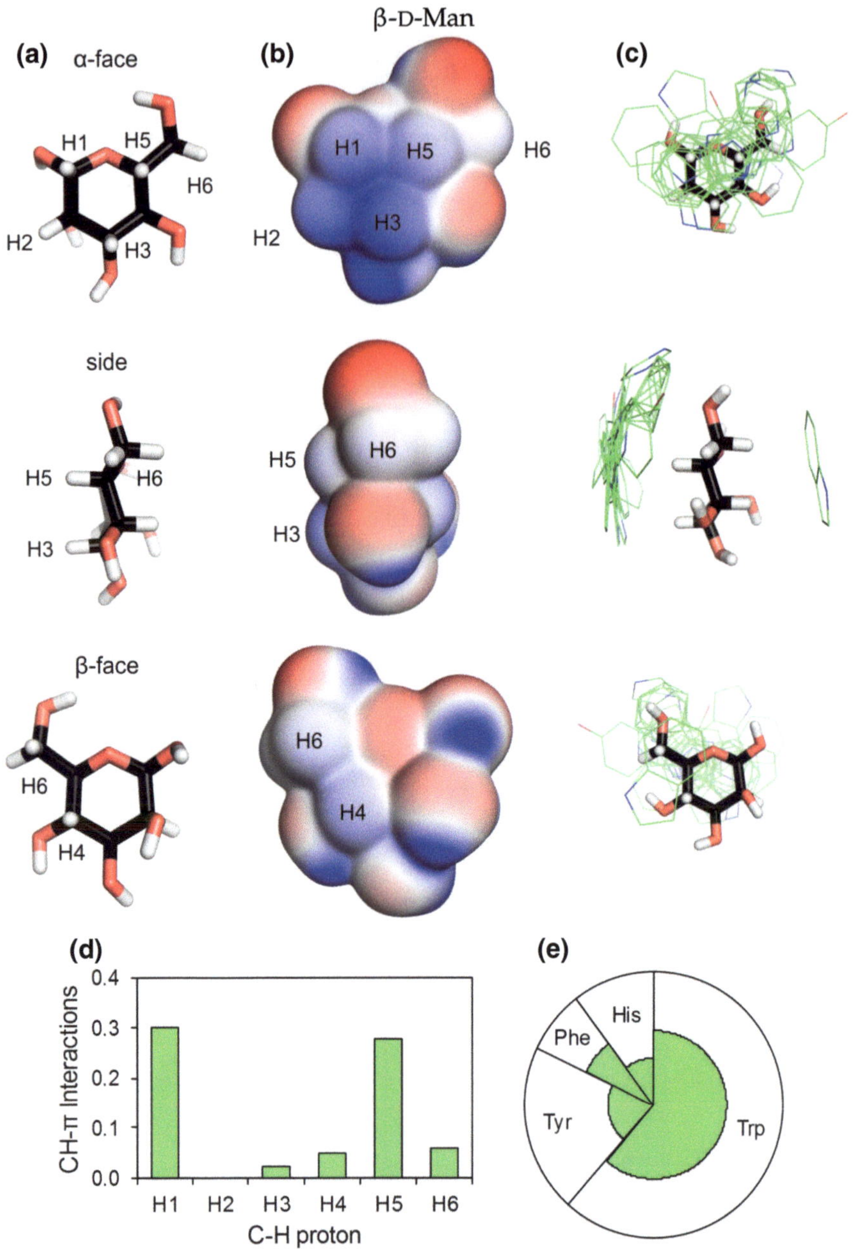

Fig. 7.7 CH?π interactions of β-D-Man in data set. **a** Three views of optimised structure. **b** ESP calculated for optimised structure. **c** All monosaccharide residues in data set overlaid, with identified amino acids CH?π acceptor aromatic rings shown in green. **d** Average involvement of C?H protons in CH?π interactions. **e** White pie chart: proximal aromatic amino acids; Green wedges: those involved in CH?π interactions

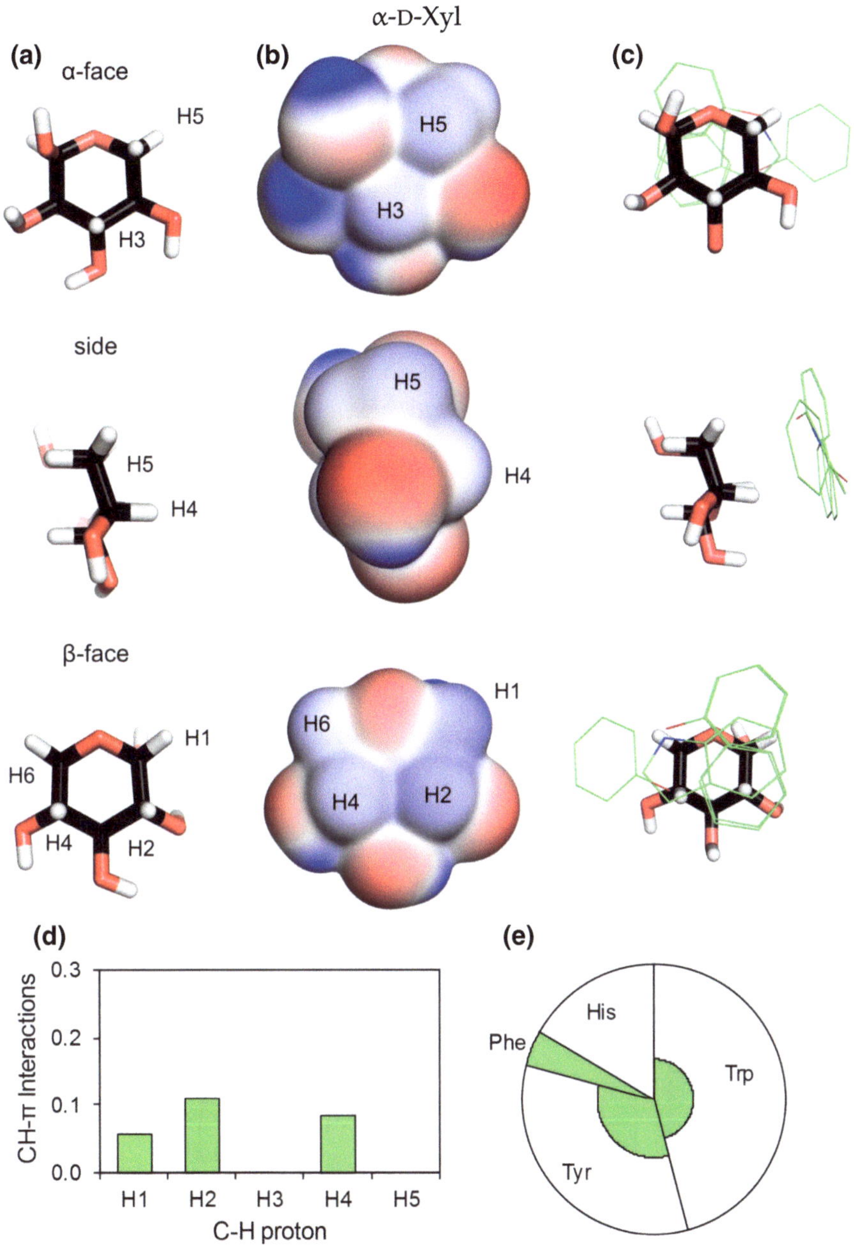

Fig. 7.8 CH?π interactions of α-D-Xyl in data set. **a** Three views of optimised structure. **b** ESP calculated for optimised structure. **c** All monosaccharide residues in data set overlaid, with identified amino acids CH?π acceptor aromatic rings shown in green. **d** Average involvement of C?H protons in CH?π interactions. **e** White pie chart: proximal aromatic amino acids; Green wedges: those involved in CH?π interactions

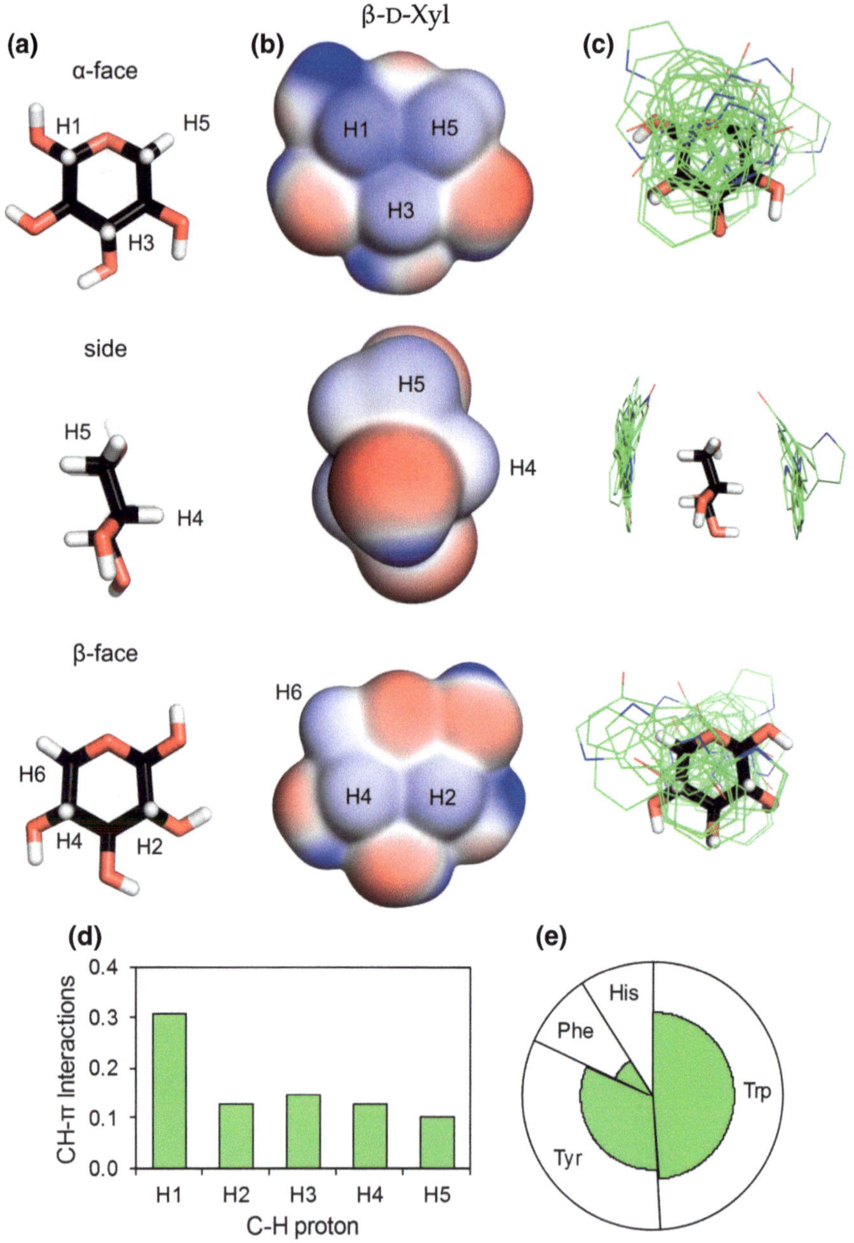

Fig. 7.9 CH?π interactions of β-D-Xyl in data set. **a** Three views of optimised structure. **b** ESP calculated for optimised structure. **c** All monosaccharide residues in data set overlaid, with identified amino acids CH?π acceptor aromatic rings shown in green. **d** Average involvement of C?H protons in CH?π interactions. **e** White pie chart: proximal aromatic amino acids; Green wedges: those involved in CH?π interactions

7.4 Peptide Synthesis and Analytical Data

Peptide Synthesis

Peptides were synthesized *via* solid-phase synthesis using a *CEM* Liberty Blue microwave peptide synthesiser. Standard Fmoc procedure was followed at either 0.1 or 0.5 mmol scale: The appropriate solid support resin (Rink Amide ChemMatrix resin for peptides with a *C*-terminal amide, HMPB ChemMatrix resin pre-loaded with the *C*-terminal amino acid for peptides with a *C*-terminal acid) was allowed to swell for 10 min in DMF before the method was started. *N*-Fmoc protected amino acids with orthogonal side-chain protection were made up as solutions in DMF. The coupling agent was diisopropylcarbodiimide (DIC) and the activator base was 1-hydroxybenzotriazole (HOBt), both made up as solutions in DMF. Deprotection of the Fmoc groups was carried out using morpholine, made up as a 20% solution by volume in DMF. After completion of the synthesis and final *N*-terminal deprotection, the resin was washed with DMF and then CH_2Cl_2 three times each. The peptide was cleaved from the resin and side chain protecting groups removed by TFA containing 5% TIPS and 5% H_2O by volume, and shaking for 3 h at RT. After this time, the resin was filtered and washed with TFA three times. The combined filtrate was reduced by evaporation and added to chilled Et_2O and stored at −20 °C for 1 h. After this time, the precipitated peptide was collected by centrifugation (4000 rpm, 4 °C, 15 min). The Et_2O was decanted and the peptide dissolved in 1:1 MeCN:H_2O and lyophilised. The crude peptide was stored at 4 °C.

Manual Peptide Coupling Conditions

Azidonorleucine (**21**)

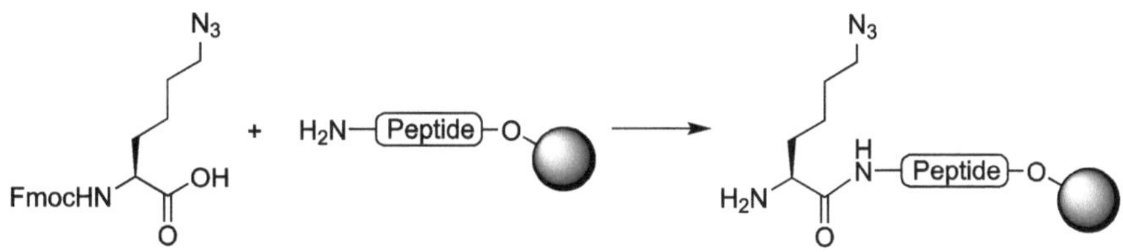

Fmoc- **21**

N-Fmoc-azidonorleucine (2 equiv. relative to peptide synthesis scale) was dissolved in DMF (5 mL per 0.1 mmol peptide). HBTU (1.9 equiv.) was added, followed by DIEA (4 equiv.), and the mixture shaken to pre-activate for 10 min. The resin was washed with DMF (3×) and then added, and the suspended resin was shaken for 3 h at RT and then washed with DMF (3×) and CH_2Cl_2 (3×). Removal of the Fmoc group was achieved by suspending the resin in 20% piperidine in DMF (5 mL per 0.1 mmol peptide), shaking at RT for 3 h and washing again before standard cleavage.

Propiolic Acid (75)

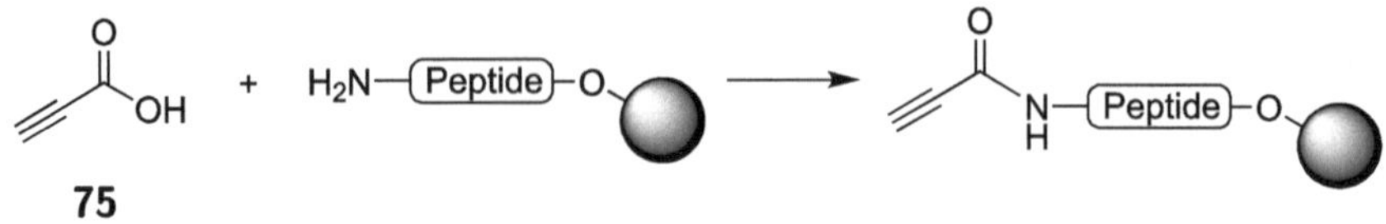

Coupling was carried out by washing the resin with *N*-methylpyrrolidone (NMP) (3×) and suspending it in NMP (5 mL per 0.1 mmol peptide). HOBt (6 equiv.) was added to the suspension, followed by propiolic acid (**75**, 5 equiv.) and then DIC (4.5 equiv.). The suspended resin was shaken for 3 h at RT and then washed with DMF (3×) and CH_2Cl_2 (3×) before cleavage as normal.

Acid-4-Alkyne (57)

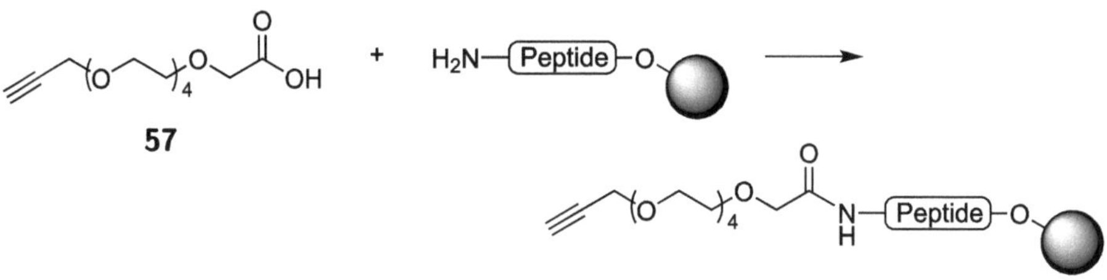

Acid-4-Alkyne (**57**, 1.2 equiv. relative to peptide) was dissolved in DMF (10 mL per 0.1 mmol peptide). HBTU (1.18 equiv.) was added, and then DIEA (2 equiv.), and the mixture shaken to pre-activate for 15 min. The resin was washed with DMF (3×) and then added, and the suspended resin was shaken for 3 h at RT and then washed with DMF (3×) and CH_2Cl_2 (3×) before standard cleavage.

Peptide Mass Spectrometry

MALDI MS was carried out using an *Applied Biosystems* 4700 Proteomics analyser (MALDI-time-of-flight (TOF)-MS) in the linear mode using a saturated solution of α-cyano-4-hydroxycinnamic acid in 1:1 H_2O:MeCN as the matrix. MS data are reported as mass/charge (M/z) values, which are equivalent to Da for species with a 1+ charge.

High-Performance Liquid Chromatography

Analytical

Analytical reversed-phase HPLC was carried out on a *JASCO* system with Pu2086 pumps fitted with a DG-2080-53 degasser, MX-2080-32 mixer and MD-910 detector using a *Phenomenex* Kinetex C18 column (100 mm × 4.6 mm, 5 μm). HPLC-grade H_2O (solvent A) and MeCN (solvent B), each containing 0.1% TFA, were used as the mobile phase at a flow rate of 1 mL min^{-1}. A linear gradient was run as indicated over 15 min unless stated otherwise. Absorbance of light at 220 nm in the UV spectrum was recorded.

Semi-Preparative

Purification by semi-preparative reversed-phase HPLC was carried out using a *JASCO* system with Pu2086 pumps fitted with a DG-2080-53 degasser, MX-2080-32 mixer and UV-2077 detector using a *Phenomenex* Luna C18 column (150 mm × 10 mm, 5 μm). HPLC-grade H_2O (solvent A) and MeCN (solvent B), each containing 0.1% TFA, were used as the mobile phase at a flow rate of 3 mLmin^{-1}. A linear gradient was run as indicated over 25 min and fractions containing the desired product were identified by MALDI MS and those containing the pure compound (by analytical HPLC) were collected and lyophilised. The pure product was stored at −20 °C.

Preparative

Purification by preparative reversed-phase HPLC was carried out on a *JASCO* system with Pu2086 pumps fitted with a DG-2080-53 degasser, MX-2080-32 mixer and UV-2077 detector using a *Vydac* 218TP C18 column (300 mm × 20 mm, 5 μm). HPLC-grade H_2O (solvent A) and MeCN (solvent B), each containing 0.1% TFA, were used as the mobile phase at a flow rate of 8 mLmin^{-1}. A linear gradient was run as indicated over 40 min and fractions containing the desired product were identified by MALDI MS and those containing the pure compound (by analytical HPLC) were collected and lyophilised. The pure product was stored at −20 °C.

Circular Dichroism Spectroscopy

CD measurements were made using a *JASCO* J-815 spectropolarimeter fitted with a Peltier temperature controller. Peptide solutions were prepared as described in MOPS buffer at pH 7.4 and examined in 1 mm *Hellma* quartz cuvettes. Spectra were recorded from 190 to 260 nm set at a 50 nm min^{-1} scan rate, 1 nm interval, 1nm bandwidth and 1 s response times over 8 accumulations. After baseline correction, recorded ellipticities in ° were converted to molar residual ellipticities (°cm^{-2} dmol^{-1} res^{-1}) by normalising for the concentration of peptide bonds and path length.

hSAF-p1

`H-KIAALKAKIAALKAEIAALEWENAALEA-OH`

MALDI MS: Calculated = 2921, Observed = 2922.

Preparative HPLC: 25–60% solvent B.

Analytical HPLC: 20–80% solvent B, retention time = 8.0 min.

CD spectra: 100 μM peptide at 20 °C (solid blue) and 37 °C (dotted red).

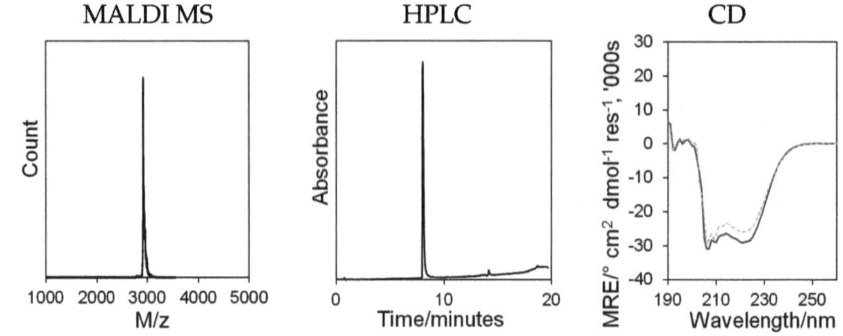

hSAF-p1-K1Z

H-ZIAALKAKIAALKAEIAALEWENAALEA-OH

MALDI MS: Calculated = 2949, Observed = 2945.

Preparative HPLC: 25–60% solvent B.

Analytical HPLC: 20–60% solvent B, retention time = 11.0 min.

CD spectra: 100 μM peptide at 20 °C (solid blue) and 37 °C (dotted red).

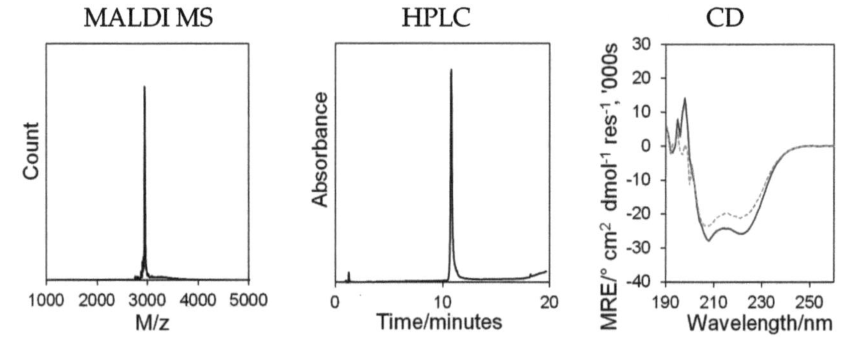

hSAF-p1-A14Z

H-KIAALKAKIAALKZEIAALEWENAALEA-OH

MALDI MS: Calculated = 3004, Observed = 2999.

Preparative HPLC: 25–60% solvent B.

Analytical HPLC: 20–60% solvent B, retention time = 12.0 min.

CD spectra: 100 μM peptide at 20 °C (solid blue) and 37 °C (dotted red).

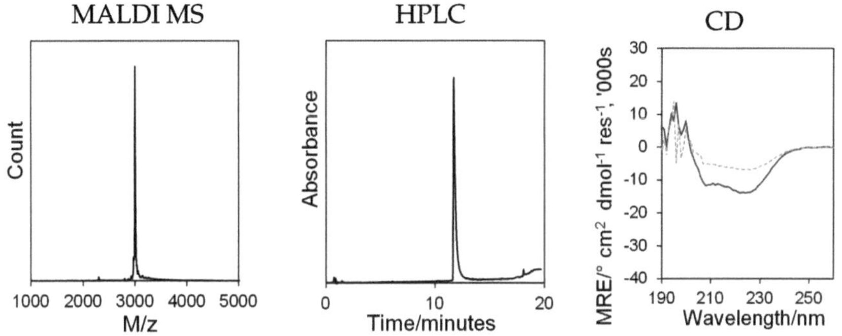

hSAF-p2

H-KIAALKAKNAALKAEIAALEWEIAALEA-OH

MALDI MS: Calculated = 2921, Observed = 2923.

Preparative HPLC: 25–60% solvent B.

Analytical HPLC: 20–60% solvent B, retention time = 11.0 min.

CD spectra: 100 μM peptide at 20 °C (solid blue) and 37 °C (dotted red).

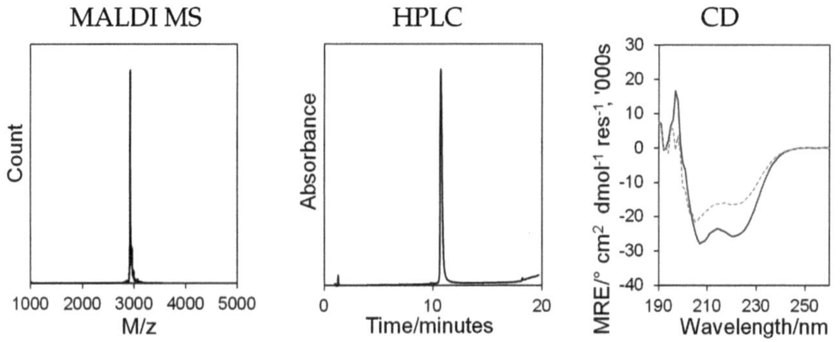

hSAF-p2sw

H-EIAALEWEIAALEAKIAALKAKNAALKA-OH

MALDI MS: Calculated = 2921, Observed = 2939.

Preparative HPLC: 25–60% solvent B.

Analytical HPLC: 20–60% solvent B, retention time = 10.5 min.

CD spectra: 100 μM peptide at 20 °C (solid blue) and 37°C (dotted red).

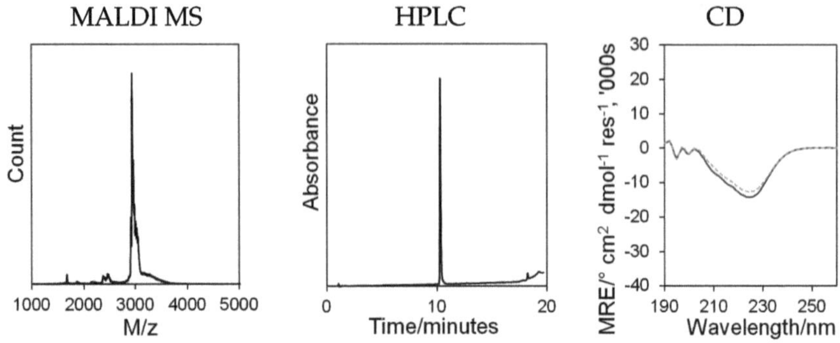

RGD-peptide

`Pr-GSGYGRGDS-NH2`, Pr = propiolic acid

MALDI MS: Calculated = 906, Observed = 899.

Preparative HPLC: 5–40% solvent B.

Analytical HPLC: 5–40% solvent B, retention time = 9.5 min.

CD spectra: 100 μM peptide at 20 °C (solid blue) and 37 °C (dotted red).

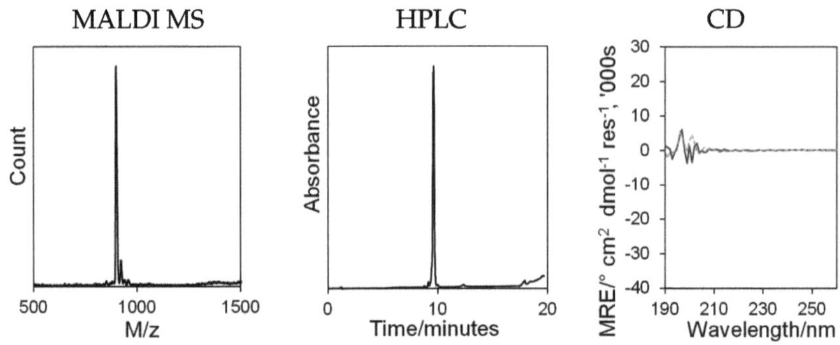

RGD-4-alkyne

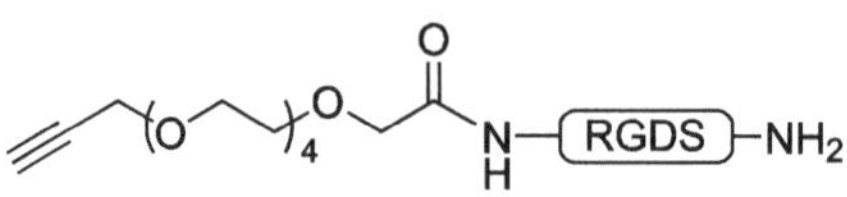

MALDI MS: Calculated = 705, Observed = 699.
Preparative HPLC: 5–40% solvent B.
Analytical HPLC: 5–40% solvent B, retention time = 8.5 min.

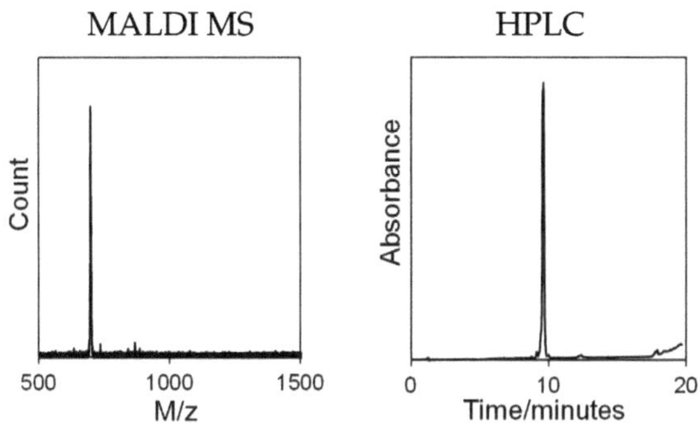

Decoration of Free Peptides

A solution was made up with azide-containing hSAF-p1 variant at 1 mm in aq. MOPS buffer at pH 7.4. The modifier was added to give a concentration of 2 mm, followed by $CuSO_4$ and then sodium ascorbate, both in solution and to give a concentration of 4 mm of each. The reaction mixture was left for 30 min at RT before purification using the same conditions as the unmodified peptide.

Decorated hSAF-p1-K1Z

hSAF-p1-K1Z+RGD-peptide

Analytical HPLC: 30–45% solvent B, retention time = 6.5 min.

CD spectra: 100 μM peptide at 20 °C (solid blue) and 37 °C (dotted red).

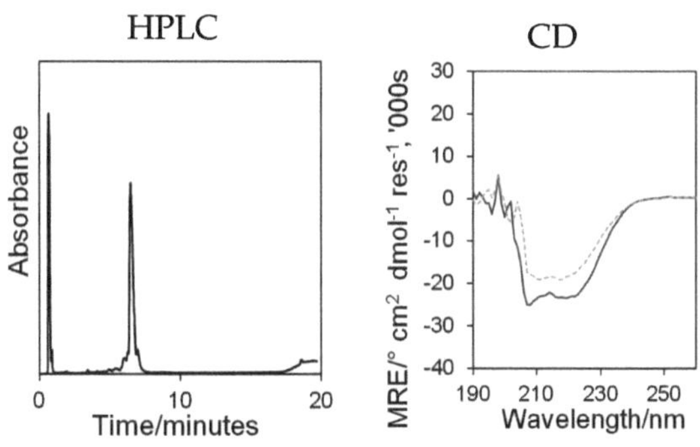

hSAF-p1-K1Z+Glc-0

MALDI MS: Calculated = 3167, Observed = 3162.

Analytical HPLC: 25–60% solvent B, retention time = 10.5 min.

CD spectra: 100 μM peptide at 20 °C (solid blue) and 37 °C (dotted red).

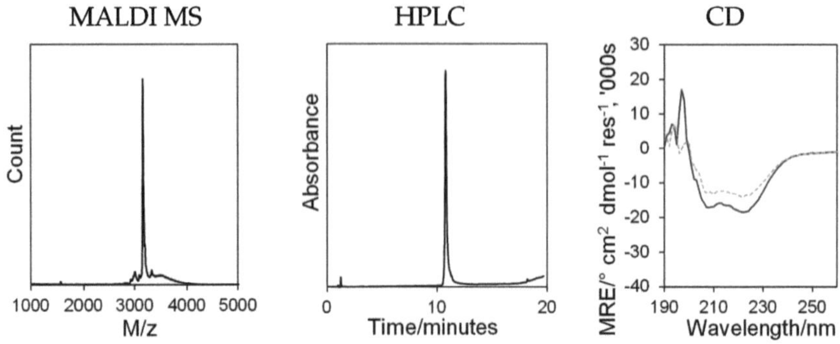

hSAF-p1-K1Z+Gal-0

MALDI MS: Calculated = 3167, Observed = 3164.

Analytical HPLC: 30–50% solvent B, retention time = 7.0 min.

CD spectra: 100 μM peptide at 20 °C (solid blue) and 37 °C (dotted red).

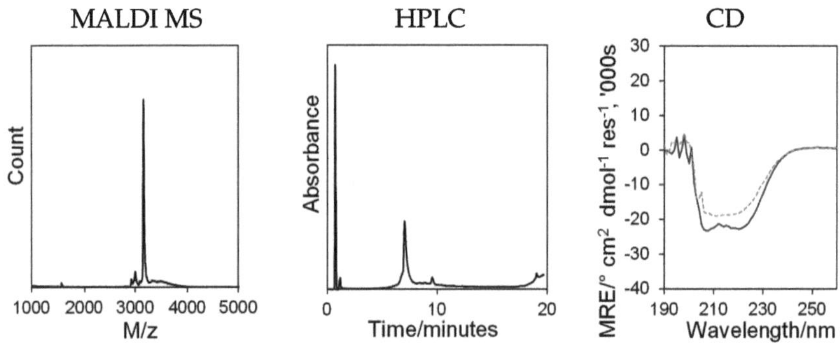

hSAF-p1-K1Z+GlcNAc-0

MALDI MS: Calculated = 3207, Observed = 3202.

Analytical HPLC: 30–50% solvent B, retention time = 7.0 min.

CD spectra: 100 μM peptide at 20 °C (solid blue) and 37 °C (dotted red).

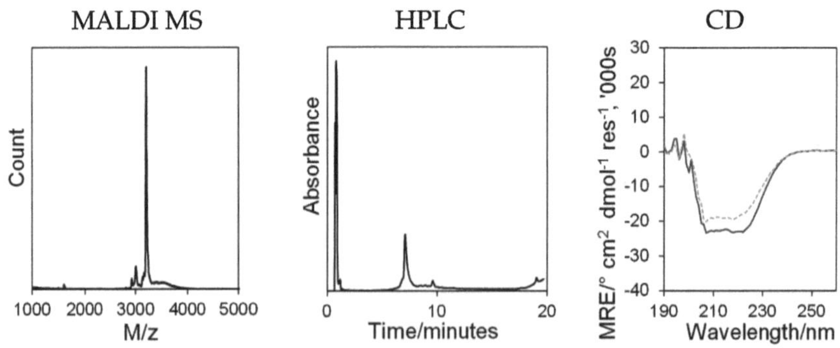

Decorated hSAF-p1-A14Z

hSAF-p1-A14Z+Glc-0

MALDI MS: Calculated = 3222, Observed = 3223.

Analytical HPLC: 20–60% solvent B, retention time = 10.0 min.

CD spectra: 100 μM peptide at 20 °C (solid blue) and 37 °C (dotted red).

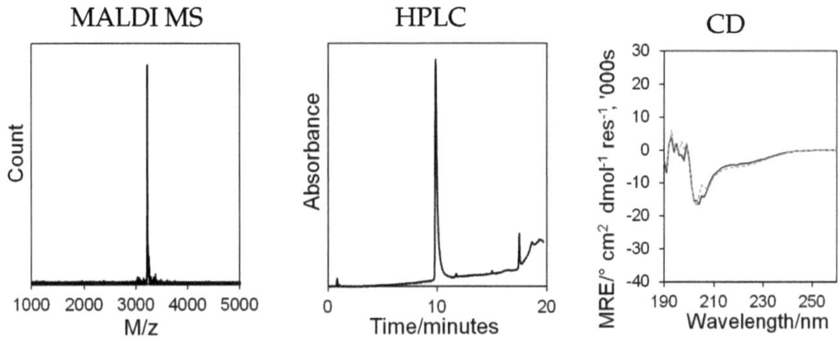

hSAF-p1-A14Z+Gal-0

MALDI MS: Calculated = 3222, Observed = 3225.

Analytical HPLC: 20–60% solvent B, retention time = 10.0 min.

CD spectra: 100 μM peptide at 20 °C (solid blue) and 37 °C (dotted red).

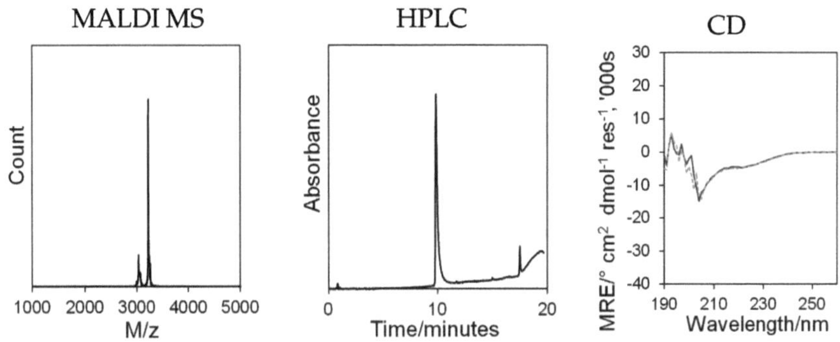

hSAF-p1-A14Z+GlcNAc-0

MALDI MS: Calculated = 3262, Observed = 3262.

Analytical HPLC: 20–60% solvent B, retention time = 10.0 min.

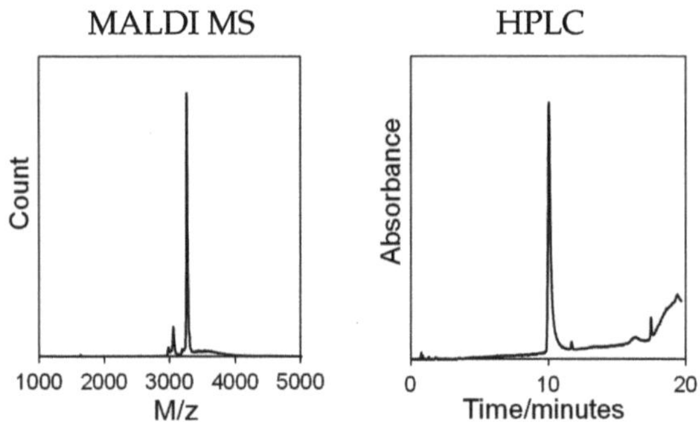

hSAF-p1-A14Z+SGal-0

MALDI MS: Calculated = 3302, Observed = 3226 (-SO_3H).

Analytical HPLC: 20–60% solvent B over 9 min, retention time = 7.5 min.

CD spectrum: 100 μM peptide at 20 °C (solid blue).

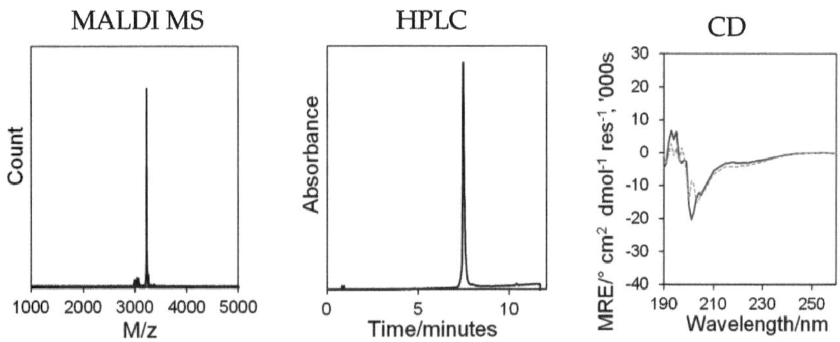

hSAF-p1-A14Z+RGD-peptide

MALDI MS: Calculated = 3910, Observed = 3907.

CD spectra: 100 μM peptide at 20 °C (solid blue) and 37 °C (dotted red).

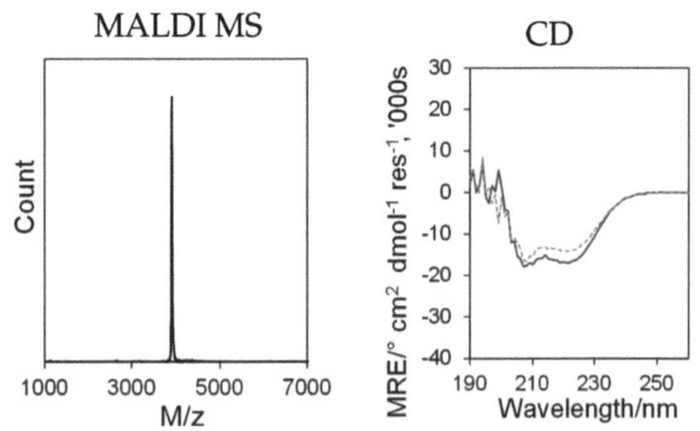

7.5 Peptide Fibres, Heterodimers, and Hydrogels

Peptide Manipulations

Water purified with a *Merck Millipore* Milli-Q Integral Water Purification System was used for peptide manipulations. MOPS buffer was prepared as a 10× solution to give a final (1×) concentration of MOPS at 40 mm, sodium acetate at 10 mm, and EDTA at 1 mm. PBS buffer contained NaCl at 137 mm, KCl at 2.7 mm, Na_2HPO_4 at 10 mm, and KH_2PO_4 at 1.8 mm. Both were corrected to pH 7.4 using HCl and/or NaOH.

7.5.1 Peptide Fibres

A solution was made up with hSAF-p1 (or a hSAF-p1 variant) and hSAF-p2 in a 1:1 ratio with a total final concentration of 50 μM in each peptide in aq. MOPS buffer at pH 7.4.

For post-formation CuAAC decoration, the modifier was added to give a final concentration of 100 μM, followed by $CuSO_4$ and then sodium ascorbate, both in solution and to give a final concentration of 200 μM of each. The reaction mixture was left for 30 min at RT before analysis.

CD Spectra of Fibres

Measured at 20 °C (solid blue) and 37 °C (dotted red).

Parent hSAFs—hSAF-p1 & hSAF-p2

Also themal unfolding (temperature dependence of MRE_{222}) and CD spectra at 55 °C (solid yellow) and 5 °C after annealing (dotted purple).

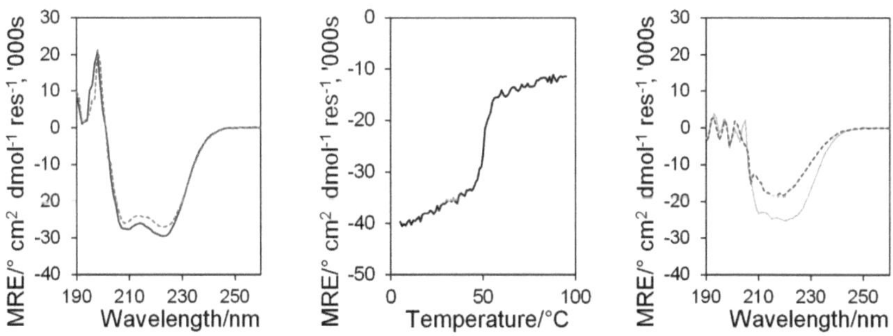

Modifiable hSAFs—hSAF-p1-K1Z & hSAF-p2

Also thermal unfolding (temperature dependence of MRE_{222}) and CD spectra at 55 °C (solid yellow) and 5 °C after annealing (dotted purple).

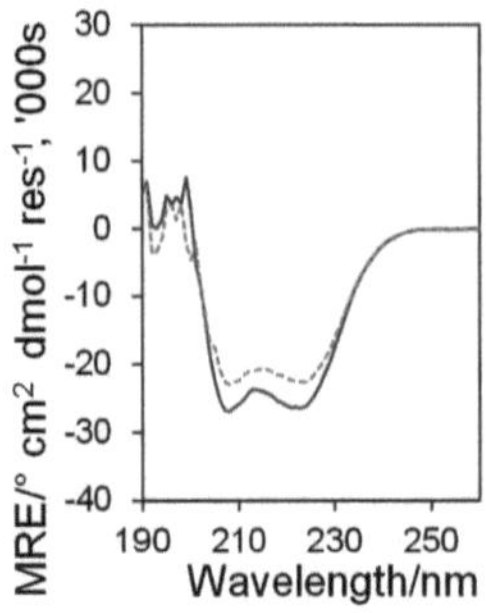

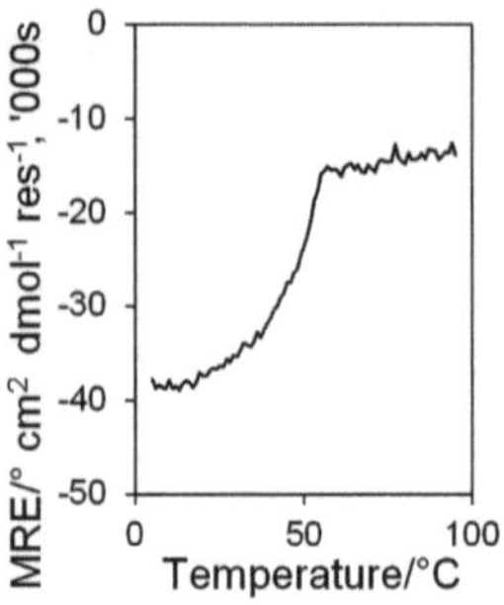

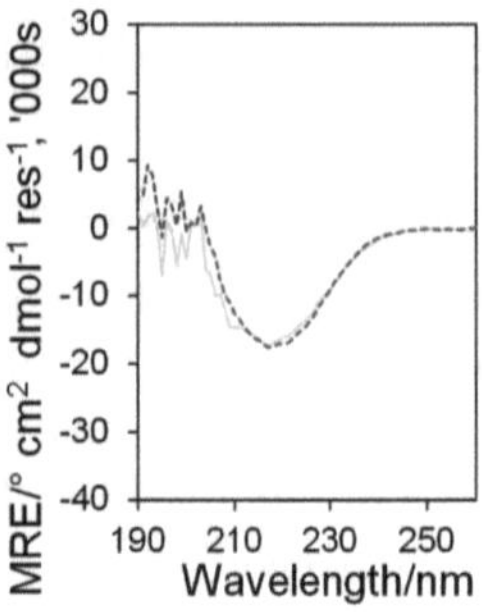

Decorated hSAF-p1-K1Z hSAFs

hSAF-p1-K1Z & hSAF-p2 + Cu

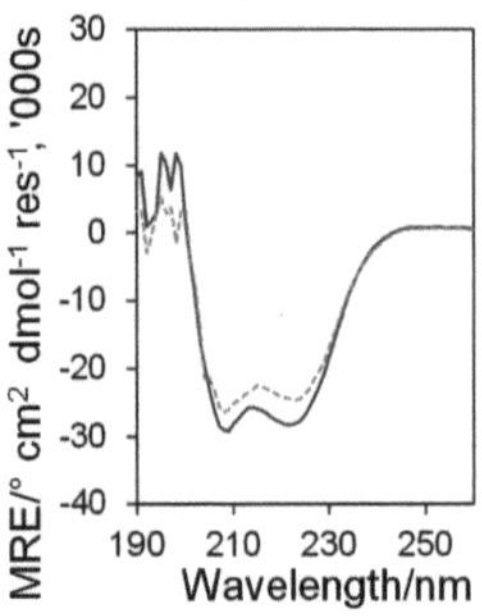

hSAF-p1-K1Z+RGD-0 & hSAF-p2

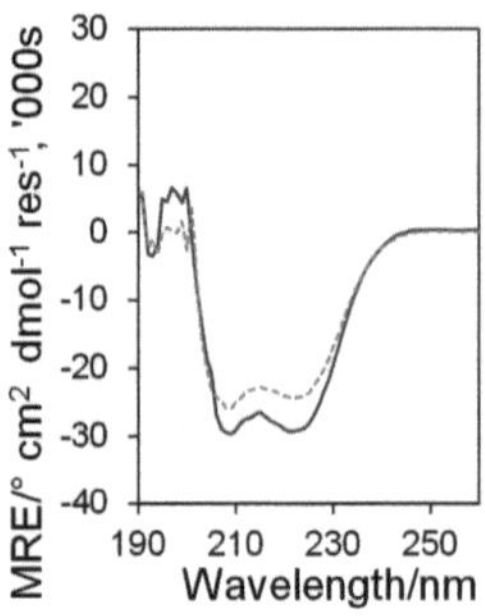

hSAF-p1-K1Z+Glc-0 & hSAF-p2

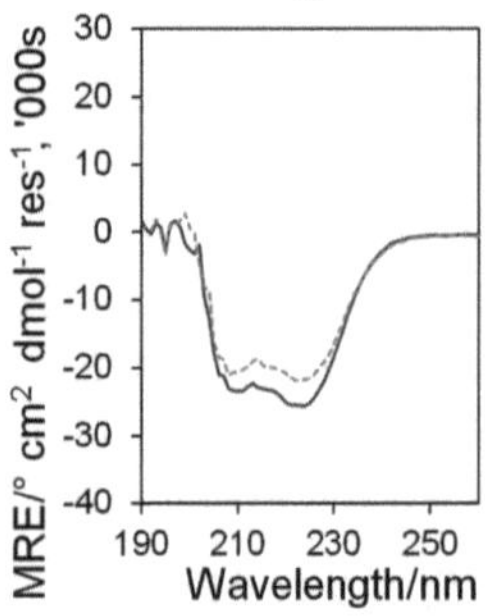

Modifiable hSAFs—hSAF-p1-A14Z and hSAF-p2

hSAF-p1-A14Z & hSAF-p2

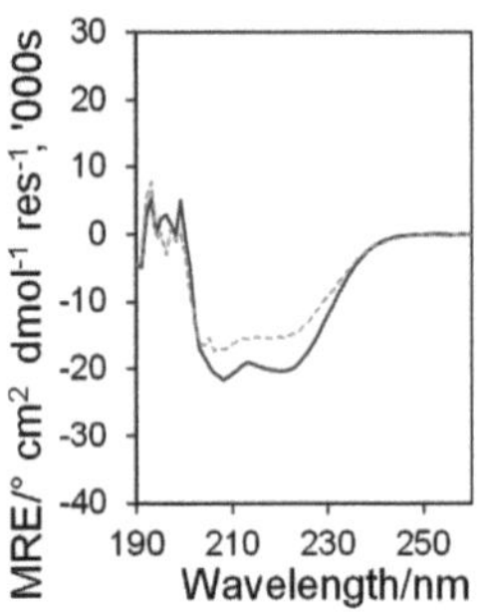

hSAF-p1-A14Z+Glc-0 & hSAF-p2

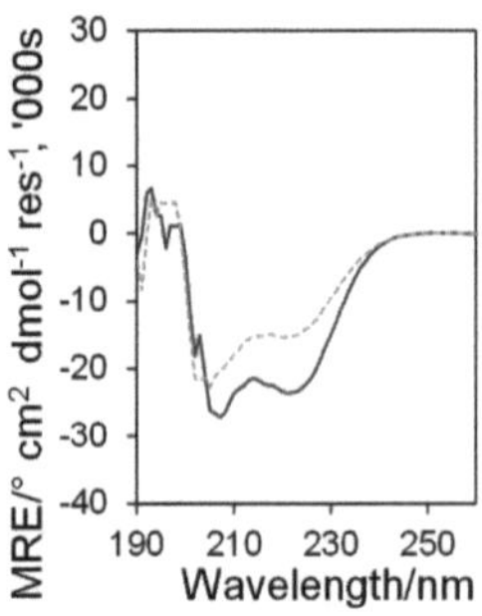

hSAF-p1-A14Z+Gal-0 & hSAF-p2

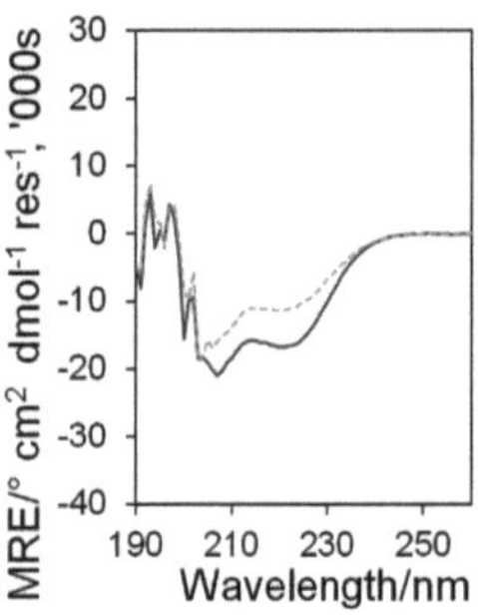

hSAF-p1-A14Z+SGal-0
& hSAF-p2

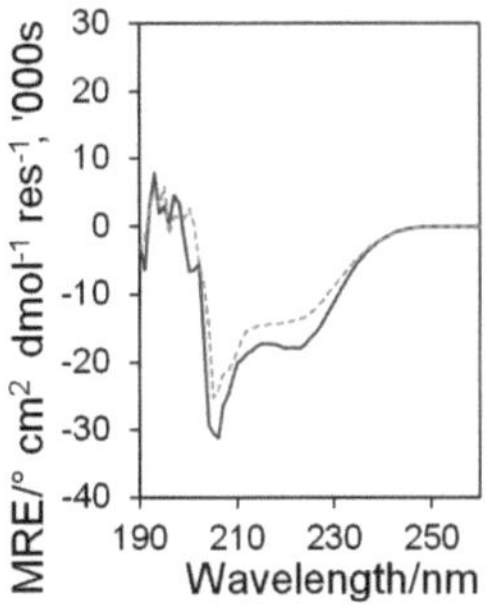

TEM of Fibres

Samples were prepared with hSAF peptides at a total concentration of 100 μM in aq. MOPS buffer at pH 7.4. TEM grids were prepared by spotting 6 μL of the sample onto a carbon-coated copper grid. This was left for 60 s and the supernatant removed. The grid was washed with H_2O (6 μL) and then immediately stained with 1% w/v aq. uranyl acetate, which was left for 30 s and then removed. The fibres were visualised with a *JEOL* JEM 1200 EX MKII Transmission Electron Microscope. Images were recorded digitally with a Mega View II digital camera, using *Soft Imaging Systems GmbH* Analysis 3.0 image analysis software.

Decorated hSAF-p1-K1Z hSAFs

hSAF-p1-K1Z+Glc-0 & hSAF-p2

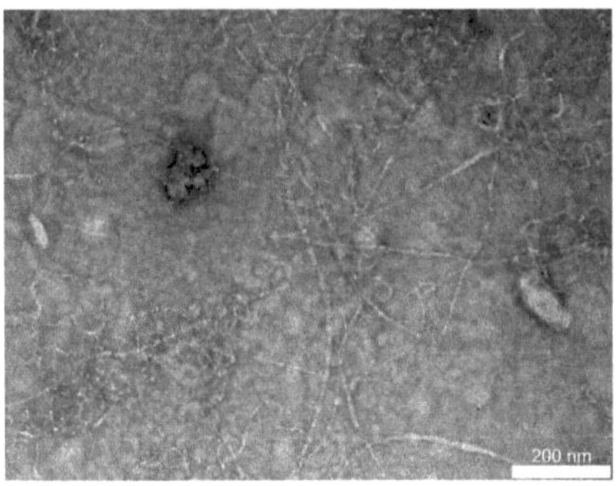

hSAF-p1-K1Z+Gal-0 & hSAF-p2

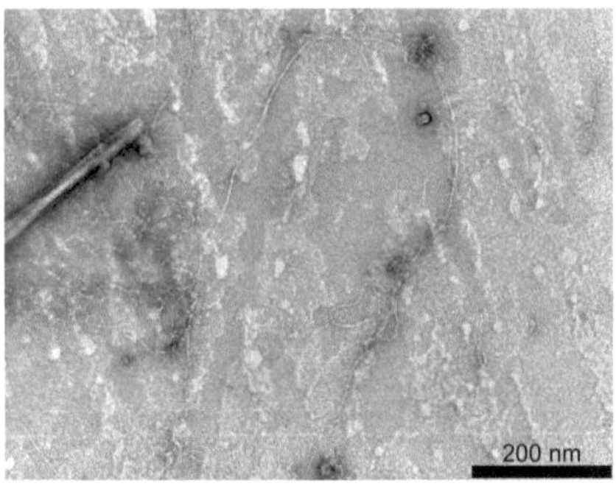

hSAF-p1-K1Z+GlcNAc-0 & hSAF-p2

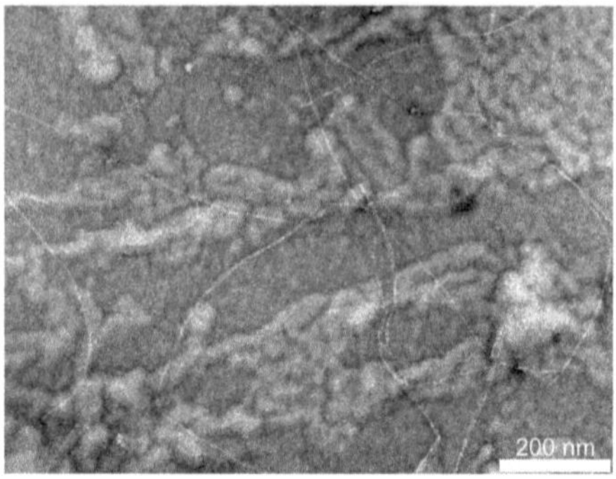

7.5.2 *Discrete Heterodimers*

A solution was made up with hSAF-p1 (or a hSAF-p1 variant) and hSAF-p2sw in a 1:1 ratio with a total final concentration of 50 μM in each peptide in aq. MOPS buffer at pH 7.4.

For post-formation CuAAC decoration, the modifier was added to give a final concentration of 100 μM, followed by $CuSO_4$ and then sodium ascorbate, both in solution and to give a final concentration of 200 μM of each. The reaction mixture was left for 30 min at RT before analysis.

CD Spectra of Heterodimers

Measured at 20 °C (solid blue) and 37 °C (dotted red).

hSAF-p1 & hSAF-p2sw

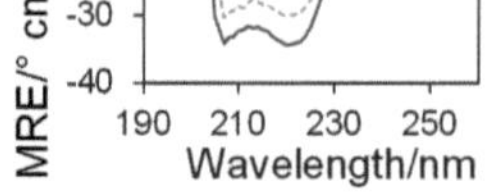

hSAF-p1-K1Z & hSAF-p2sw

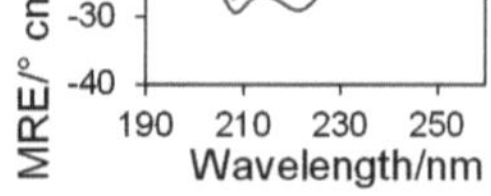

hSAF-p1-A14Z & hSAF-p2sw

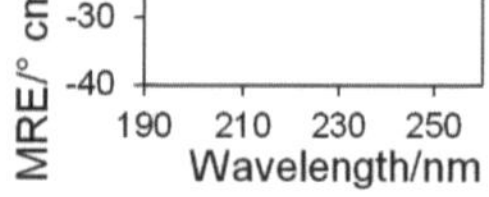

Analytical Ultracentrifugation of Heterodimers

Acknowledgement: This data was obtained by Antony Burton.

Sedimentation equilibrium experiments were conducted at 20 °C in a *Beckman-Optima* XL-I analytical ultracentrifuge fitted with an An-60 Ti rotor. Peptide solutions were prepared at 75 μM total hSAF concentration in aq. MOPS buffer at pH 7.4

and spun at speeds in the range 43 000–56 000 rpm. Data were fitted simultaneously assuming a single ideal species model using Ultrascan [13]. The partial specific volumes of the peptide and the solvent density (1.005 g mL^{-1}) were calculated using Sednterp [14].

hSAF-p1 & hSAF-p2sw – theoretical dimer mass = 5842, results fit single species of mass 5796.

Raw data at measured rpm values, with residuals to fit below:

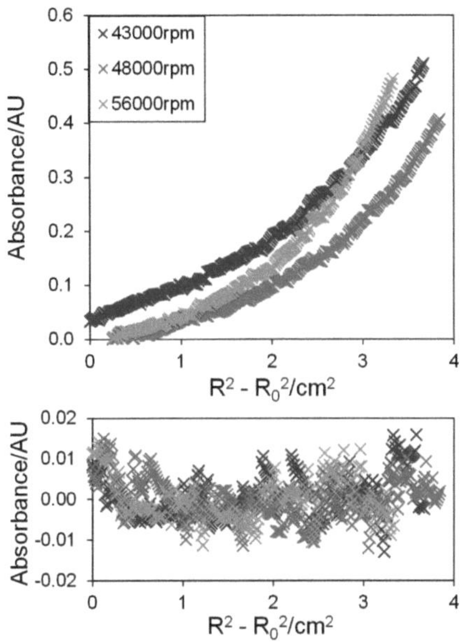

7.5.3 hSAF Hydrogels

All constituents—H_2O, 10× MOPS buffer, hSAF-p1 (or variant) and hSAF-p2 solutions—were cooled on ice before use. The gel was made up in a vessel cooled on ice in aq. MOPS buffer at pH 7.4 with each peptide at a final concentrations of 1 mM. The vessel was left on ice for 30 min, and then the gel was left overnight at RT.

For post-formation CuAAC decoration (for gels formed with hSAF-p1-K1Z), the next day a solution of equal volume to the hydrogel with the modifier at 2 mM and $CuSO_4$ and sodium ascorbate both at 4 mM was added. The gel was then left at RT overnight, and the next morning the supernatant removed and the gel washed with aq. EDTA (10 mM, 4× gel volume, 3 repeats) and aq. PBS buffer at pH 7.4 (4× gel volume, 3 repeats), left for 20 min for each washing step, and finally left under PBS buffer.

Rheology

Viscoelastic properties were measured on a *Bohlin* CVO rotational rheometer fitted with a 20 mm conical plate with a cone angle of 4°. Gels were formed in situ on the plate and left at 5 °C for 30 min prior to measurement at the specified temperature.

Frequency response of G′ (circles) and G″ (diamonds) of hydrogels at 20 °C (blue) and 37 °C (red).

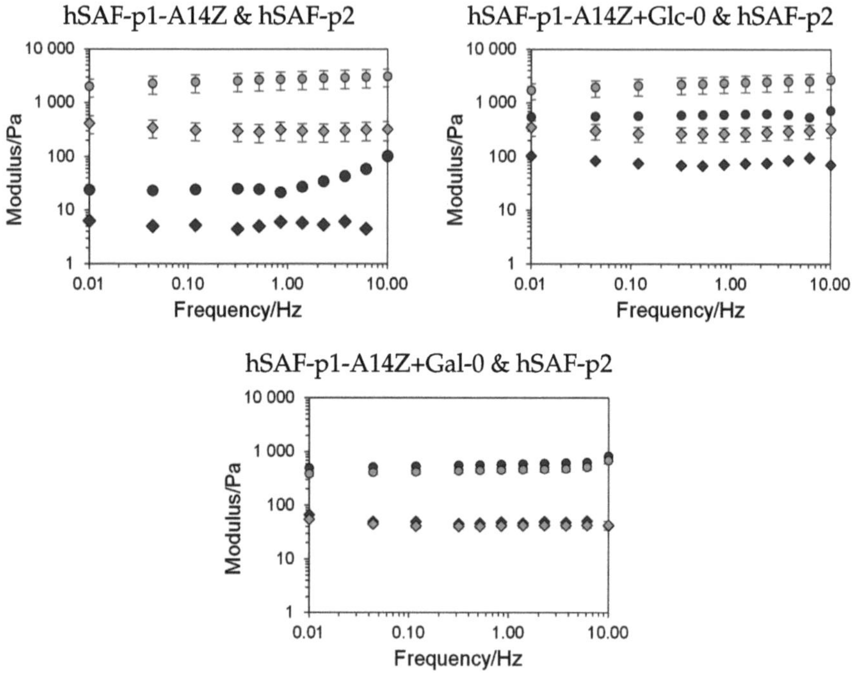

Copper Assay of Decorated Gels

Three reagents were prepared: reagent A consisted of 30 g trichloroacetic acid dissolved in 100 mL H_2O, reagent B consisted of 8.8 mg L-gulonic acid in 25 mL H_2O, and reagent C consisted of 6 mg BCA, 3.6 g NaOH and 15.6 g anhydr. HEPES in 90 mL H_2O. 60 μL of the sample was diluted with H_2O (25 μL) and then reagent A (25 μL) was added. This was vortexed and centrifuged (10 000 rpm, 5 min) and then 50 μL of the supernatant was added to reagent B (10 μL), followed by reagent C (40 μL). The mixture was vortexed and left overnight. The UV-Vis spectrum was then measured on a *PerkinElmer* Lambda 25 UV/Vis spectrophotometer at wavelength 355 nm, measured against a blank from a sample containing equivalent peptides but no copper.

Calibration curve (calibrated with solutions of known concentration of $CuSO_2$).

UV absorbance spectra of unwashed hSAF hydrogel (dotted red line), showing characteristic absorbance at 355 nm indicative of presence of copper, and washed hydrogel (solid blue line). Absorbance at 355 nm for hSAF-p1-K1Z gel decorated with RGD-peptide and washed by standard procedure: −0.006555.

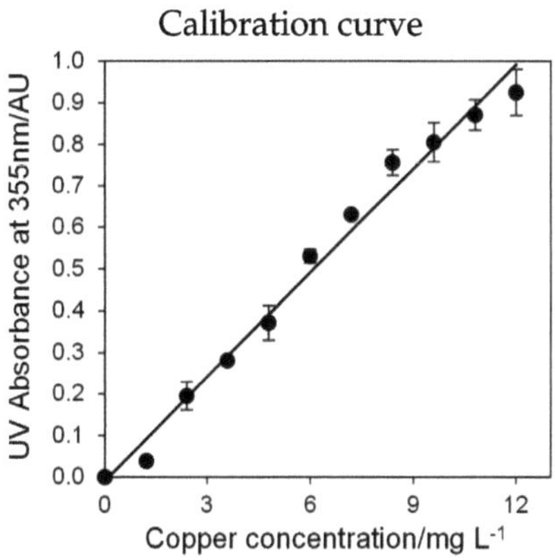

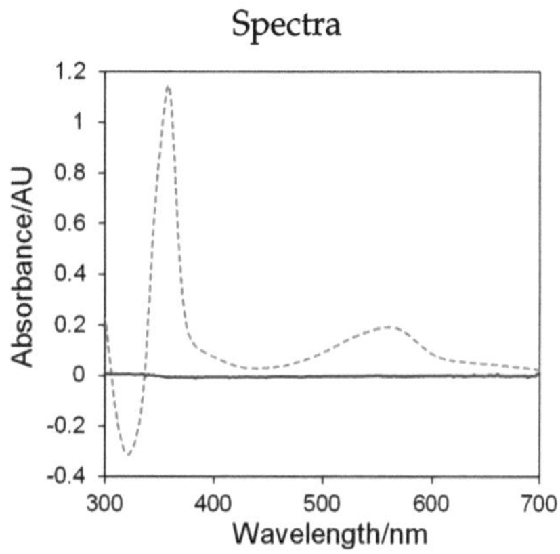

7.6 Chemical Synthesis

General Procedures

All reactions that were carried out under anhydrous conditions used flame-dried apparatus under an atmosphere of N_2. Anhydrous CH_2Cl_2 and THF were distilled HPLC-grade and further dried and degassed using a commercially available solvent purification system (*Anhydrous Engineering*). Unless stated otherwise, all other solvents and reagents were used as supplied without further purification. TLC was carried out using *Merck* aluminium-backed sheets coated with $60F_{254}$ silica gel. Visualization of silica plates was achieved with solutions of $KMnO_4$, ninhydrin, or vanillin as indicated. Flash column chromatography (FCC) was carried out using technical grade 60 Å 230–400 mesh silica gel (*Sigma-Aldrich*). 'Hexane' refers to *n*-hexane.

Amberlite IRA120 (H^+) resin was conditioned as follows: 100 g of the commercial resin was placed in a 500 mL sintered filter funnel and allowed to swell with 200 mL of acetone for 5 min. The solvent was removed by suction and the resin was washed successively with 800 mL of acetone, 500 mL methanol, 500 mL 5 M HCl and then 1 L of water or until the pH of filtrate was ~7, as indicated by pH paper. The resin was partially dried on the filter and then stored and used as needed.

Proton NMR (^{1}H NMR) spectra were recorded on a *Varian* 400-MR (400 MHz) spectrometer. Carbon NMR (^{13}C NMR) spectra were recorded on a *Varian* 400-MR (101 MHz) spectrometer. All chemical shifts are quoted on the δ scale in ppm using residual solvent as the internal standard. Coupling constants (*J*) are reported with the following splitting abbreviations or combinations thereof: s = singlet, d = doublet, t = triplet, q = quartet and m = multiplet.

Small molecule MS was carried out either by MALDI-TOF-MS ('MALDI') on an *Applied Biosystems* 4700 Proteomics analyser or by electrospray ionisation (ESI) on a *Bruker* Daltonics Microtof II.

Melting points (MPs) were measured on a *Gallenkamp* melting point apparatus and are uncorrected.

IR spectroscopy was carried out on a *Perkin-Elmer* Spectrum One FT-IR spectrophotometer. υ_{max} values are reported in cm^{-1}.

7.6.1 Representative Synthesis of Peracetylated Monosaccharide

1,2,3,4,6-Penta-*O*-acetyl-D-glucopyranose (76)

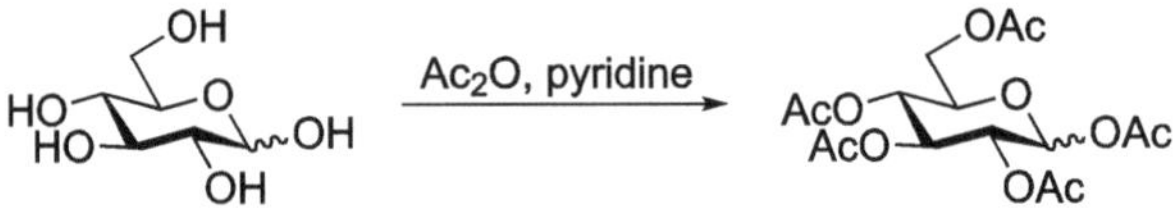

23 **76**

D-Glucose (**23**, 10 g, 56 mmol) in a 250 mL three-necked round-bottomed flask equipped with a Teflon-coated magnetic stirrer was dried briefly under reduced pressure. Freshly distilled pyridine (40 mL, 520 mmol) was added and the solution stirred at RT. Acetic anhydride (50 mL, 450 mmol) was added dropwise over 30 min and the reaction mixture stirred at RT for 16 h. After this time, excess reagents were removed under reduced pressure to give an oil. Ethanol was added and the product left to crystallise, before filtering and washing with ice-cold ethanol. The resultant solid was recrystallised from 1:1 EtOAc:hexane to give a colourless crystalline solid (19.6 g, 90% yield) as a ratio of anomers (α:β = 4:1, determined by NMR, major anomer characterised). Published compound [15]. **TLC** $R_f = 0.3$ (1:1 EtOAc:hexane, vanillin). ^{1}**H NMR** (400 MHz, $CDCl_3$) $\delta_H = 2.00$ (s, 3H, $COCH_3$), 2.01 (s, 3H, $COCH_3$), 2.02 (s, 3H, $COCH_3$), 2.08 (s, 3H, $COCH_3$), 2.16 (s, 3H, $COCH_3$), 4.04–4.13 (m, 2H, H5 & H6), 4.24 (m, 1H, H6′), 5.05–5.16 (m, 2H, H2 & H4), 5.46 (m, 1H, H3), 6.31 (d, J = 3.5 h, 1H, H1). ^{13}**C NMR** (100 MHz, $CDCl_3$) $\delta_C = 20.54$ ($CO\underline{C}H_3$), 20.66 ($CO\underline{C}H_3$), 20.76 ($CO\underline{C}H_3$), 20.79 ($CO\underline{C}H_3$), 20.97 ($CO\underline{C}H_3$), 61.51 (C6), 67.93, 69.24 (C2, C3 & C4), 69.88 (C5), 89.13 (C1), 168.85 ($\underline{C}OCH_3$), 169.48 ($\underline{C}OCH_3$), 169.75 ($\underline{C}OCH_3$), 170.32 ($\underline{C}OCH_3$), 170.73 ($\underline{C}OCH_3$).

7.6.2 Synthesis of Simple Alkynyl Linkers

Acknowledgement: Some of these syntheses and characterisations were carried out with Ian Hazledon.

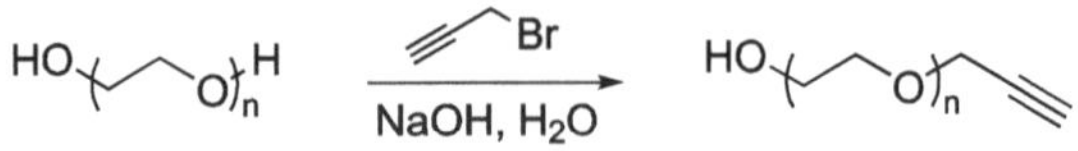

KOH was placed in a round-bottomed flask equipped with a Teflon-coated magnetic stirrer, followed by H_2O. The solution was stirred and the specified ethylene glycol was added and the solution stirred for 10 min. The solution was then cooled to 0 °C with an ice bath and 3-bromo-1-propyne (80% w/w in toluene) was added and the reaction mixture stirred for 5 min, after which the ice bath was removed and stirring continued at room temperature for 16 h. After this time the reaction mixture was carefully neutralised with 3 M aq. HCl before being extracted with EtOAc (3× equal volume to H_2O). The combined EtOAc was washed with brine (equal volume to H_2O), dried over Na_2SO_4, filtered and the solvent removed under reduced pressure. The resultant residue was purified by FCC (EtOAc) to give the product.

2-Prop-2-ynoxyethanol (37, n = 1)

4.3 g KOH (76 mmol) in 10 mL H_2O, 10.6 mL ethylene glycol (**34**, 180 mmol), 3 mL 3-bromo-1-propyne (38 mmol). Product obtained as pale yellow liquid (2.2 g, 55% yield). Published compound [16]. **TLC** $R_f = 0.55$ (EtOAc, $KMnO_4$). **IR** (oil) υ_{max} = 3402 (m), 3288 (m), 2934 (m), 2866 (m), 2115 (w), 1106 (s), 1066 (s), 1027 (s). ^{1}**H NMR** (400 MHz, $CDCl_3$) δ_H = 2.48 (t, J = 2.5 Hz, 1H, C≡CH), 3.65–3.69 (m, 2H, $CH_2\underline{CH_2}OC$), 3.76–3.80 (m, 2H, $\underline{CH_2}OH$), 4.22 (d, J = 2.5 Hz, 2H, $OCH_2C\equiv$). ^{13}**C NMR** (100 MHz, $CDCl_3$) δ_C = 58.47 ($O\underline{C}H_2C\equiv$), 61.63 (CH_2OH), 71.24 ($CH_2\underline{C}H_2OC$), 74.92 ($CH_2\underline{C}\equiv CH$), 79.46 ($C\equiv\underline{C}H$). **EA** Calculated = C 59.98, H 8.05. Found = C 59.98, H 7.71.

Propargyl diethylene glycol (2-(2-(prop-2-yn-1-yloxy)ethoxy)ethan-1-ol, 38, n = 2)

900 mg KOH (16 mmol) in 3 mL H_2O, 3.8 mL diethylene glycol (**35**, 40 mmol), 800 μL 3-bromo-1-propyne (8 mmol). Product obtained as clear liquid (790 mg, 69% yield). Published compound [17]. **TLC** $R_f = 0.35$ (EtOAc, $KMnO_4$). **IR** (oil) υ_{max} = 3402 (m), 3257 (m), 2870 (m), 2114 (w), 1368 (s), 1093 (s), 1063 (s), 1031 (s). ^{1}**H NMR** (400 MHz, $CDCl_3$) δ_H = 2.43 (t, J = 2.5 Hz, 1H, C≡CH), 3.56–3.61 (m, 2H, CH_2OH), 3.64–3.74 (m, 6H, 3× $CH_2\underline{CH_2}OC$), 4.19 (d, J = 2.5 Hz, 2H, $OCH_2C\equiv$). ^{13}**C NMR** (101 MHz, $CDCl_3$) δ_C = 58.40 ($O\underline{C}H_2C\equiv$), 61.71 ($CH_2\underline{C}H_2OC$), 69.11 ($CH_2\underline{C}H_2OC$), 70.20 ($CH_2\underline{C}H_2OC$), 72.47 (CH_2OH), 74.69 ($CH_2\underline{C}\equiv CH$), 79.43 ($C\equiv\underline{C}H$). **MS** (ESI) m/z $[M + Na]^+$ Calculated = 167.0679. Found = 167.0679.

Propargyl tetraethylene glycol (3,6,9,12-Tetraoxapentadec-14-yn-1-ol, 39, n = 4)

1.6 g KOH (28 mmol) in 4 mL H_2O, 15 mL tetraethylene glycol (**36**, 70 mmol), 1.4 mL 3-bromo-1-propyne (14 mmol). Product obtained as pale yellow liquid (2.3 g, 71% yield). Published compound [18]. **TLC** $R_f = 0.35$ (9:1 Et_2O:MeOH, $KMnO_4$). ^{1}**H NMR** (400 MHz, $CDCl_3$) δ_H = 2.42 (t, J = 2.5 Hz, 1H, C≡CH), 3.52–3.56 (m, 2H, CH_2OH), 3.64–3.74 (m, 14H, 7× $CH_2\underline{CH_2}OC$), 4.19 (d, J = 2.5 Hz, 2H, $OCH_2C\equiv$). ^{13}**C NMR** (101 MHz, $CDCl_3$) δ_C = 58.83 ($O\underline{C}H_2C\equiv$), 61.73 ($CH_2\underline{C}H_2OC$), 69.09 ($CH_2\underline{C}H_2OC$), 70.33 ($CH_2\underline{C}H_2OC$), 70.39 ($CH_2\underline{C}H_2OC$), 70.53 ($CH_2\underline{C}H_2OC$), 70.56 ($CH_2\underline{C}H_2OC$), 70.61 ($CH_2\underline{C}H_2OC$), 72.48 (CH_2OH),

74.50 ($C{\equiv}\underline{C}H$), 79.61 ($CH_2\underline{C}{\equiv}CH$). **MS** (ESI) m/z $[M + Na]^+$ Calculated = 255.1203. Found = 255.1199.

7.6.3 *Synthesis of Alkynyl Gluco- and Galactosides with Simple Linkers*

Acknowledgement: Some of these syntheses and characterisations were carried out with Ian Hazledon.

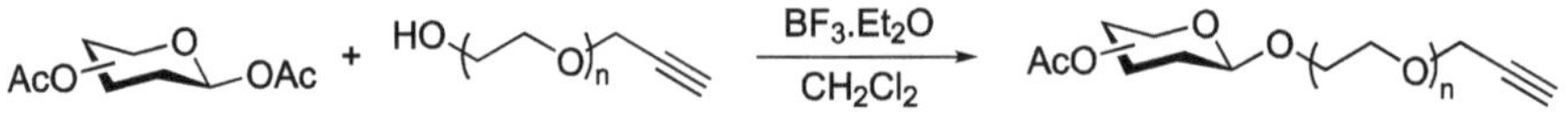

The specified 1,2,3,4,6-penta-*O*-acetyl-β-D-pyranose (1 equivs.) was placed in a three-necked round-bottomed flask equipped with a Teflon-coated magnetic stirrer and dried briefly under reduced pressure. Anhydr. CH_2Cl_2 (20 mL per 1 g pyranose) was added and the solution cooled to 0 °C with an ice bath. The specified alkynyl alcohol (1.5 equivs.) was added followed by $BF_3.Et_2O$ (1.5 equivs.) and the reaction mixture was stirred for 5 min. The ice bath was removed, and stirring continued at RT for 16 h. K_2CO_3 (480 mg per 1 g pyranose) was added and the reaction mixture stirred for a further 30 min. The resultant suspension was then filtered and the solid washed with H_2O (equal volume to CH_2Cl_2). The CH_2Cl_2 was separated and the H_2O washed with CH_2Cl_2 (2× equal volume to H_2O). The combined CH_2Cl_2 was dried over Na_2SO_4, filtered and the solvent removed under reduced pressure to yield a pale yellow solid. This was purified by FCC (1:1 EtOAc:hexane) to give the product.

PerAcGlc-0-Alkyne (1-*O*-Prop-2-ynyl-2,3,4,6-tetra-*O*-acetyl-β-D-glucopyranoside, 77)

1,2,3,4,6-Penta-*O*-acetyl-β-D-glucopyranose (**30**, 10 g, 26 mmol) with 2-propyn-1-ol. Product a colourless crystalline solid (8.90 g, 90% yield). Published compound [19]. **TLC** $R_f = 0.35$ (1:1 EtOAc:hexane, vanillin). **MP** = 114 °C (Et_2O-hexane) [Lit. [19] = 114 − 115 °C (CH_2Cl_2-hexane)]. **IR** (crystal) υ_{max} = 3275 (m), 2971 (m), 2875 (m), 2580 (m), 1753 (s), 1732 (s), 1233 (s), 1208 (s), 1038 (s). **^{1}H NMR** (400 MHz, $CDCl_3$) δ_H = 2.00 (s, 3H, $COCH_3$), 2.02 (s, 3H, $COCH_3$), 2.05 (s, 3H, $COCH_3$), 2.08 (s, 3H, $COCH_3$), 2.46 (t, J = 2.5 Hz, 1H, $C{\equiv}CH$), 3.72 (ddd, J = 10.0 Hz, 4.5 Hz, 2.5 Hz, 1H, H5), 4.14 (dd, J = 12.5 Hz, 3.5 Hz, 1H, H6), 4.27 (dd, J = 12.5 Hz, 4.5 Hz, 1H, H6′), 4.37 (d, J = 2.5 Hz, 2H, $OCH_2C{\equiv}$), 4.77 (d, J = 8.0 Hz, 1H, H1), 5.01 (dd, J = 9.5 Hz, 8.0 Hz, 1H, H2), 5.10 (dd, J = 10.5 Hz, 9.5 Hz, 1H, H4), 5.24 (dd, J = 10.5 Hz, 10.5 Hz, 1H, H3). **^{13}C NMR** (100 MHz, $CDCl_3$) δ_C = 20.57 ($CO\underline{C}H_3$), 20.59 ($CO\underline{C}H_3$), 20.67 ($CO\underline{C}H_3$), 20.71 ($CO\underline{C}H_3$), 55.91 ($O\underline{C}H_2C{\equiv}$), 61.74 (C6), 68.27 (C4), 70.93 (C2), 71.90 (C5), 72.74 (C3), 75.46 ($C{\equiv}\underline{C}H$), 78.07 ($CH_2\underline{C}{\equiv}CH$), 98.09 (C1), 169.37 ($\underline{C}OCH_3$), 169.41 ($\underline{C}OCH_3$), 170.23 ($\underline{C}OCH_3$), 170.63 ($\underline{C}OCH_3$). **MS** (ESI) $[M + Na]^+$ Calculated = 409.1105. Found = 409.1101. **EA** Calculated = C 52.85, H 5.74. Found = C 53.11, H 5.98.

PerAcGlc-2-Alkyne (1-*O*-(2-(prop-2-yn-1-yloxy)ethoxy)ethan-1-yl-2,3,4,6-tetra-*O*-acetyl-β-D-glucopyranoside, 40)

1,2,3,4,6-Penta-*O*-acetyl-β-D-glucopyranose (**30**, 390 mg, 1 mmol) with propargyl diethylene glycol (**38**). Product a colourless crystalline solid (280 mg, 60% yield). Novel compound. **TLC** $R_f = 0.30$ (3:2 EtOAc:hexane, $KMnO_4$). **IR** (film) $\upsilon_{max} =$ 3276 (m), 2945 (m), 2877 (m), 2117 (w), 1744 (s), 1366 (s), 1213 (s), 1031 (s). **^{1}H NMR** (400 MHz, $CDCl_3$) $\delta_H = 2.00$ (s, 3H, $COCH_3$), 2.01 (s, 3H, $COCH_3$), 2.04 (s, 3H, $COCH_3$), 2.08 (s, 3H, $COCH_3$), 2.44 (t, $J = 2.5$ Hz, 1H, C≡CH), 3.58–3.80 (m, 8H, 3× $CH_2\underline{CH_2}OC$ & H5 & $CH_2\underline{CH_2}OC1$), 3.94 (ddd, $J = 11.0$ Hz, 4.5 Hz, 4.0 Hz, 1H, $CH_2\underline{CH'_2OC}1$), 4.13 (dd, $J = 12.5$ Hz, 2.5 Hz, 1H, H6), 4.20 (d, $J = 2.5$ Hz, 2H, $OCH_2\underline{C}$≡), 4.25 (dd, $J = 12.5$ Hz, 4.5 Hz, 1H, H6′), 4.62 (d, $J = 8.0$ Hz, 1H, H1), 4.99 (dd, $J = 9.5$ Hz, 8.0 Hz, 1H, H2), 5.08 (dd, $J = 10.0$ Hz, 9.5 Hz, 1H, H4), 5.20 (dd, $J = 9.5$ Hz, 9.5 Hz, 1H, H3). **^{13}C NMR** (100 MHz, $CDCl_3$) $\delta_C = 20.58$ ($CO\underline{C}H_3$), 20.61 ($CO\underline{C}H_3$), 20.67 ($CO\underline{C}H_3$), 20.73 ($CO\underline{C}H_3$), 58.44 ($O\underline{C}H_2C$≡), 61.93 (C6), 68.40 (C4), 68.89 (4× $CH_2\underline{C}H_2OC$), 71.19 (C2), 71.81 (C5), 72.78 (C3), 74.59 (C≡$\underline{C}$H), 79.45 ($CH_2\underline{C}$≡CH), 100.90 (C1), 169.38 ($\underline{C}OCH_3$), 169.39 ($\underline{C}OCH_3$), 170.26 ($\underline{C}OCH_3$), 170.66 ($\underline{C}OCH_3$). **MS** (ESI) $[M + Na]^+$ Calculated = 497.1629. Found = 497.1619.

PerAcGlc-4-Alkyne (1-*O*-3,6,9,12-Tetraoxapentadec-14-yn-1-yl-2,3,4,6-tetra-*O*-acetyl-β-D-glucopyranoside, 41)

1,2,3,4,6-Penta-*O*-acetyl-β-D-glucopyranose (**30**, 500 mg, 1.3 mmol) with propargyl tetraethylene glycol (**39**). Product a colourless oil (430 mg, 60% yield). Published compound [20]. **TLC** $R_f = 0.10$ (1:1 EtOAc:hexane, $KMnO_4$). **IR** (oil) $\upsilon_{max} =$ 3274 (m), 2941 (m), 2872 (m), 2114 (w), 1745 (s), 1366 (s), 1214 (s), 1032 (s). **^{1}H NMR** (400 MHz, $CDCl_3$) $\delta_H = 1.99$ (s, 3H, $COCH_3$), 2.01 (s, 3H, $COCH_3$), 2.03 (s, 3H, $COCH_3$), 2.07 (s, 3H, $COCH_3$), 2.43 (t, $J = 2.5$ Hz, 1H, C≡CH), 3.57–3.78 (m, 16H, 7× $CH_2\underline{CH_2}OC$ & H5 & $CH_2\underline{CH_2}OC1$), 3.93 (dd, $J = 11.0$ Hz, 4.0 Hz, 1H, H6), 4.13 (m, 1H, $CH_2\underline{CH'_2}OC1$), 4.19 (d, $J = 2.5$ Hz, 2H, OCH_2C≡), 4.25 (dd, $J =$ 12.5 Hz, 4.5 Hz, 1H, H6′), 4.60 (d, $J = 8.0$ Hz, 1H, H1), 4.98 (dd, $J = 9.5$ Hz, 8.0 Hz, 1H, H2), 5.07 (dd, $J = 9.5$ Hz, 9.5 Hz, 1H, H4), 5.19 (dd, $J = 9.5$ Hz, 9.5 Hz, 1H, H3). **^{13}C NMR** (100 MHz, $CDCl_3$) $\delta_C = 20.58$ ($CO\underline{C}H_3$), 20.60 ($CO\underline{C}H_3$), 20.66 ($CO\underline{C}H_3$), 20.73 ($CO\underline{C}H_3$), 58.38 ($O\underline{C}H_2C$≡), 61.94 (C6), 68.40 (C4), 69.07, 69.09, 70.27, 70.38, 70.55, 70.59, 70.68 (8× $CH_2\underline{C}H_2OC$), 71.25 (C2), 71.76 (C5), 72.82 (C3), 74.50 (C≡$\underline{C}$H), 79.65 ($CH_2\underline{C}$≡CH), 100.83 (C1), 169.33 ($\underline{C}OCH_3$), 169.37 ($\underline{C}OCH_3$), 170.24 ($\underline{C}OCH_3$), 170.64 ($\underline{C}OCH_3$). **MS** (ESI) $[M + Na]^+$ Calculated = 585.2154. Found = 585.2143.

PerAcGal-0-Alkyne (1-*O*-Prop-2-ynyl-2,3,4,6-tetra-*O*-acetyl-β-D-galactopyranoside, 78)

1,2,3,4,6-Penta-*O*-acetyl-β-D-galactopyranose (**31**, 500 mg, 1.3 mmol) with 2-propyn-1-ol. Product a colourless crystalline solid (360 mg, 74% yield). Published compound [21]. **TLC** $R_f = 0.35$ (1:1 EtOAc:hexane, vanillin). **MP** = 73°C (Et_2O-hexane). **IR** (crystal) $\upsilon_{max} =$ 3299 (m), 3262 (m), 2943 (m), 1742 (s), 1368 (s),

1225 (s), 1045 (s). **^{1}H NMR** (400 MHz, $CDCl_3$) δ_H = 1.99 (s, 3H, $COCH_3$), 2.05 (s, 3H, $COCH_3$), 2.07 (s, 3H, $COCH_3$), 2.15 (s, 3H, $COCH_3$), 2.46 (t, J = 2.5 Hz, 1H, $C{\equiv}CH$), 3.93 (ddd, J = 6.5 Hz, 6.5 Hz, 1.0 Hz, 1H, H5), 4.13 (dd, J = 11.0 Hz, 6.5 Hz, 1H, H6), 4.18 (dd, J = 11.0 Hz, 6.5 Hz, 1H, H6′), 4.38 (d, J = 2.5 Hz, 2H, $OCH_2C{\equiv}$), 4.73 (d, J = 8.0 Hz, 1H, H1), 5.06 (dd, J = 10.0 Hz, 3.5 Hz, 1H, H3), 5.22 (dd, J = 11.0 Hz, 8.0 Hz, 1H, H2), 5.40 (dd, J = 3.5 Hz, 1.0 Hz, 1H, H4). **^{13}C NMR** (100 MHz, $CDCl_3$) δ_C = 20.56 ($CO\underline{C}H_3$), 20.63 ($CO\underline{C}H_3$), 20.65 ($CO\underline{C}H_3$), 20.77 ($CO\underline{C}H_3$), 55.88 ($O\underline{C}H_2C{\equiv}$), 61.18 (C6), 66.96 (C4), 68.47 (C2), 70.80, 70.84 (C3 & C5), 75.34 ($C{\equiv}\underline{C}H$), 78.17 ($CH_2\underline{C}{\equiv}CH$), 98.63 (C1), 169.54 ($\underline{C}OCH_3$), 170.11 ($\underline{C}OCH_3$), 170.21 ($\underline{C}OCH_3$), 170.37 ($\underline{C}OCH_3$).

PerAcGal-2-Alkyne (1-*O*-(2-(prop-2-yn-1-yloxy)ethoxy)ethan-1-yl-2,3,4,6-tetra-*O*-acetyl-β-D-galactopyranoside, 42)

1,2,3,4,6-Penta-*O*-acetyl-β-D-galactopyranose (**31**, 390 mg, 1 mmol) with propargyl diethylene glycol (**38**). Product a colourless crystalline solid (370 mg, 79% yield). Novel compound. **MP** = 68°C (Et_2O-hexane). **IR** (film) υ_{max} = 3274 (m), 2941 (m), 2874 (m), 2114 (w), 1741 (s), 1368 (s), 1214 (s), 1039 (s). **^{1}H NMR** (400 MHz, $CDCl_3$) δ_H = 1.97 (s, 3H, $COCH_3$), 2.04 (s, 3H, $COCH_3$), 2.06 (s, 3H, $COCH_3$), 2.16 (s, 3H, $COCH_3$), 2.44 (t, J = 2.5 Hz, 1H, $C{\equiv}CH$), 3.60–3.70 (m, 6H, 3× $CH_2\underline{CH_2}OC$), 3.76 (ddd, J = 11.0 Hz, 7.0 Hz, 4.0 Hz, 1H, $CH_2\underline{CH_2}OC1$), 3.88 – 3.99 (m, 2H, H5 & $CH_2\underline{CH'_2}OC1$), 4.06–4.18 (m, 2H, H6), 4.20 (d, J = 2.5 Hz, 2H, $OCH_2C{\equiv}$), 4.58 (d, J = 8.0 Hz, 1H, H1), 5.01 (dd, J = 10.5 Hz, 3.5 Hz, 1H, H3), 5.20 (dd, J = 10.5 Hz, 8.0 Hz, 1H, H2), 5.38 (dd, J = 3.5 Hz, 1.0 Hz, 1H, H4). **^{13}C NMR** (100 MHz, $CDCl_3$) δ_C = 20.56 ($CO\underline{C}H_3$), 20.64 ($CO\underline{C}H_3$), 20.66 ($CO\underline{C}H_3$), 20.75 ($CO\underline{C}H_3$), 58.37 ($O\underline{C}H_2C{\equiv}$), 61.29 (C6), 67.06 (C4), 68.80 (C2), 68.97, 69.06, 70.37, 70.48, 70.61 (C5 & 4× $CH_2\underline{C}H_2OC$), 70.91 (C3), 74.42 ($C{\equiv}\underline{C}H$), 79.56 ($CH_2\underline{C}{\equiv}CH$), 101.29 (C1), 169.47 ($\underline{C}OCH_3$), 170.12 ($\underline{C}OCH_3$), 170.23 ($\underline{C}OCH_3$), 170.35 ($\underline{C}OCH_3$). **MS** (ESI) $[M + Na]^+$ Calculated = 497.1629. Found = 497.1619.

PerAcGal-4-Alkyne (1-*O*-(3,6,9,12-Tetraoxapentadec-14-yn-1-yl)-2,3,4,6-tetra-*O*-acetyl-β-D-galactopyranoside, 43)

1,2,3,4,6-Penta-*O*-acetyl-β-D-glalactopyranose (**31**, 500 mg, 1.3 mmol) with propargyl tetraethylene glycol (**39**). Product a colourless oil (550 mg, 75% yield). Novel compound. **TLC** R_f = 0.10 (1:1 EtOAc:hexane, $KMnO_4$). **IR** (oil) υ_{max} = 3270 (m), 2921 (m), 2871 (m), 2116 (w), 1743 (s), 1368 (s), 1215 (s), 1039 (s). **^{1}H NMR** (400 MHz, $CDCl_3$) δ_H = 1.97 (s, 3H, $COCH_3$), 2.04 (s, 3H, $COCH_3$), 2.05 (s, 3H, $COCH_3$), 2.14 (s, 3H, $COCH_3$), 2.42 (t, J = 2.5 Hz, 1H, $C{\equiv}CH$), 3.86–3.99 (m, 14H, 7× $CH_2\underline{CH_2}OC$), 3.86–3.99 (m, 2H, H5 & $CH_2\underline{CH_2}OC1$), 4.07–4.17 (m, 3H, $OCH_2C{\equiv}$ & $CH_2\underline{CH'_2}OC1$), 4.17–4.24 (m, 2H, H6), 4.56 (d, J = 8.0 Hz, 1H, H1), 5.00 (dd, J = 10.5 Hz, 3.5 Hz, 1H, H4), 5.19 (dd, J = 10.5 Hz, 8.0 Hz, 1H, H2), 5.37 (d, J = 2.5, 1H, H3). **^{13}C NMR** (100 MHz, $CDCl_3$) δ_C = 20.58 ($CO\underline{C}H_3$), 20.65 ($CO\underline{C}H_3$), 20.67 ($CO\underline{C}H_3$), 20.76 ($CO\underline{C}H_3$), 58.38 ($O\underline{C}H_2C{\equiv}$), 61.29 (C6), 67.06 (C3), 68.80 (C2), 69.06, 69.09, 70.27, 70.39, 70.56, 70.58, 70.60, 70.68 (8× $CH_2\underline{C}H_2OC$), 70.69 (C5), 70.90 (C4), 74.50 ($C{\equiv}\underline{C}H$), 79.65 ($CH_2\underline{C}{\equiv}CH$), 100.35

(C1), 169.45 ($\underline{C}OCH_3$), 170.12 ($\underline{C}OCH_3$), 170.23 ($\underline{C}OCH_3$), 170.36 ($\underline{C}OCH_3$). **MS** (ESI) $[M + Na]^+$ Calculated = 585.2154. Found = 585.2152.

7.6.4 Synthesis of Alkynyl N-*Acetylglucosaminosides with Simple Linkers*

Acknowledgement: Some of these syntheses and characterisations were carried out with Ian Hazledon.

PerAcGlcNAc-0-Alkyne (1-*O*-Prop-2-ynyl-2-acetamido-2-deoxy-3,4,6-tri-*O*-acetyl-β-D-glucopyranoside, 79)

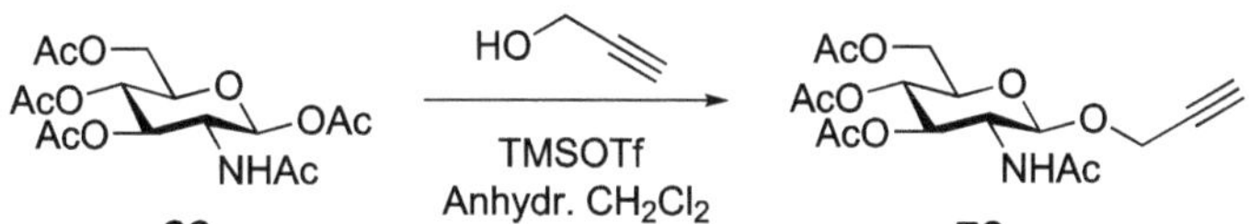

33 **79**

2-Acetamido-2-deoxy-β-D-glucopyranose-1,3,4,6-tetraacetate (**33**, 125 mg, 320 μmol) was placed in a 25 mL three-necked round-bottomed flask equipped with a Teflon-coated magnetic stirrer and dried briefly under reduced pressure. Anhydr. CH_2Cl_2 (2 mL) was added and the solution cooled to 0 °C with an ice bath. 2-Propyn-1-ol (370 μL, 6.4 mmol) was added followed by freshly distilled trimethylsilyl trifluoromethanesulfonate (65μL, 350 μmol) and the reaction mixture was stirred for 5 min. The ice bath was removed, and stirring continued at RT for a further 24 h. NEt_3 was added until the pH was >7 (pH paper) and the solvent removed under reduced pressure to yield a yellow oil. This was purified by FCC (CH_2Cl_2) to give the product as a colourless crystals (69 mg, 56% yield). Published compound [22]. **TLC** R_f = 0.75 (1:4 MeOH:CH_2Cl_2, vanillin). **MP** = 191 °C (MeOH-hexane) [Lit. [22] = 189 − 190 °C (EtOAc-hexane)]. ^{1}**H NMR** (400 MHz, $CDCl_3$) δ_H = 1.96 (s, 3H, $NHCOC\underline{H}_3$), 2.02 (s, 3H, $OCOCH_3$), 2.03 (s, 3H, $OCOCH_3$), 2.08 (s, 3H, $OCOCH_3$), 2.46 (t, J = 1.5 Hz, 1H, C≡CH), 3.72 (ddd, J = 10.0 Hz, 4.5 Hz, 2.0 Hz, 1H, H5), 3.95 (dd, J = 19.0 Hz, 9.0 Hz, 1H, H2), 4.13 (dd, J = 12.0, 2.0, 1H, H6), 4.26 (dd, J = 12.0, 4.5, 1H, H6′), 4.37 (d, J = 2.0 Hz, 2H, OCH_2C≡), 4.85 (d, J = 8.5 Hz, 1H, H1), 5.08 (dd, J = 9.5 Hz, 9.5 Hz, 1H, H4), 5.27 (dd, J = 10.0 Hz, 10.0 Hz, 1H, H3), 5.62 (d, J = 8.5 Hz, 1H, NH). ^{13}**C NMR** (100 MHz, $CDCl_3$) δ_C = 20.61 ($OCO\underline{C}H_3$), 20.69 ($OCO\underline{C}H_3$), 20.73 ($OCO\underline{C}H_3$), 23.36 ($NHCO\underline{C}H_3$), 54.25 (C2), 55.90 ($O\underline{C}H_2C$≡), 61.90 (C6), 68.41 (C4), 71.95 (C5), 72.37 (C3), 75.39 (C≡$\underline{C}$H), 78.46 ($CH_2\underline{C}$≡CH), 98.29 (C1), 169.34 ($NH\underline{C}OCH_3$), 170.37 ($O\underline{C}OCH_3$), 170.70 ($O\underline{C}OCH_3$), 170.94 ($O\underline{C}OCH_3$). **MS** (ESI) m/z $[M + Na]^+$ Calculated = 408.1265. Found = 408.1256.

Glucosamine-tetraacetate-derived oxazoline (44)

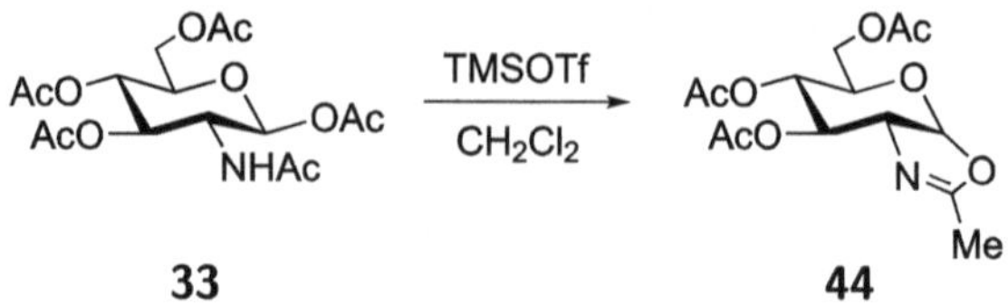

2-Acetamido-2-deoxy-β-D-glucopyranose-1,3,4,6-tetraacetate (**33**, 200 mg, 510 μmol) was placed in a 50 mL three-necked round-bottomed flask equipped with a Teflon™-coated magnetic stirrer and dried briefly under reduced pressure. Anhydr. CH_2Cl_2 (5 mL) was added and the solution cooled to 0 °C with an ice bath. Freshly distilled trimethylsilyl trifluoromethanesulfonate (100, 570 μmol) was added and the reaction mixture was stirred for 5 min. The ice bath was removed, and the reaction mixture heated to reflux and stirring continued for a further 12 h. After this the reaction mixture was cooled to room temperature and NEt_3 was added until the pH was >7 (pH paper) and the solvent removed under reduced pressure to yield a yellow oil. This was purified by FCC (49:1 CH_2Cl_2:MeOH) to give the product as a colourless liquid in quantitative yield (169 mg). Published compound [23]. **TLC** $R_f = 0.05$ (1:4 MeOH:CH_2Cl_2, ninhydrin). **^{1}H NMR** (400 MHz, $CDCl_3$) $\delta_H = 2.03$ (s, 3H, CH_3), 2.04 – 2.06 (m, 6H, 2× CH_3), 2.07 (s, 3H, CH_3), 3.56 (dt, $J = 9.0$ Hz, 4.5 Hz, 1H, H5), 4.09 (ddd, $J = 5.5$ Hz, 2.5 Hz, 1.5 Hz, 1H, H2), 4.11 – 4.14 (m, 2H, H6), 4.87 (ddd, $J = 9.0$, 2.0, 1.5, 1H, H4), 5.21 (dd, $J = 2.5$ Hz, 2.5 Hz, 1H, H3), 5.92 (d, $J = 7.5$ Hz, 1H, H1). **^{13}C NMR** (100 MHz, $CDCl_3$) $\delta_C = 13.90$ (NHCO$\underline{C}H_3$), 20.69 (OCO$\underline{C}H_3$), 20.79 (OCO$\underline{C}H_3$), 20.86 (OCO$\underline{C}H_3$), 63.28 (C6), 64.87 (C2), 67.46 (C5), 68.31 (C4), 70.28 (C3), 99.33 (C1), 166.62 (NH$\underline{C}OCH_3$), 169.14 (O$\underline{C}OCH_3$), 169.47 (O$\underline{C}OCH_3$), 170.52 (O$\underline{C}OCH_3$). **MS** (ESI) m/z $[M + Na]^+$ Calculated = 352.1003. Found = 352.1018.

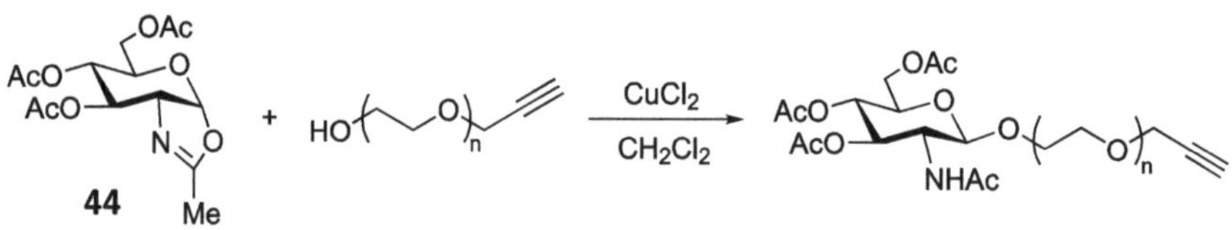

Glucosamine-tetraacetate-derived oxazoline (**44**, 1 equivs.) was placed in a round-bottomed flask and co-evaporated twice with toluene (2 mL per 100 mg oxazoline). $CuCl_2$ (1.5 equivs.) was added and the mixture co-evaporated with toluene twice more. Anhydrous CH_2Cl_2 (5 mL per 100 mg oxazoline) was added and the mixture stirred, followed by the specified alkynyl alcohol (2 equivs.). The reaction mixture was heated to reflux and stirred for 16 h. After this time the reaction mixture was cooled to room temperature and the solvent removed under reduced pressure to give a yellow oil. This was taken up in EtOAc (10 mL per 100 mg oxazoline) and washed with aq. 1 M HCl (2× 10 mL per 100 mg oxazoline), aq. saturated $NaHCO_3$ (10 mL per 100 mg oxazoline) and brine (10 mL per 100 mg oxazoline). Each aq. solution was extracted with EtOAc (10 mL per 100 mg oxazoline), and the combined organics were dried over Na_2SO_4, filtered, and the solvent removed under reduced pressure to give an oil. This was purified by FCC (19:1 Et_2O:MeOH) to give the product.

PerAcGlcNAc-1-Alkyne (1-(2-Prop-2-ynoxy)ethanyl)-2-acetamido-2-deoxy-3,4, 6-tri-*O*-acetyl-β-D-glucopyranoside, 80, n = 1)

100 mg glucosamine tetraacetate-derived oxazoline with 2-prop-2-ynoxyethanol (**37**). Product isolated as a colourless solid (78 mg, 61% yield). Novel compound. **TLC**: R_f = 0.20 (EtOAc, ninhydrin). **^{1}H NMR** (400 MHz, $CDCl_3$) δ_H = 1.95 (s, 3H, $NHCOC\underline{H}_3$), 2.01 (s, 3H, $OCOCH_3$), 2.02 (s, 3H, $OCOCH_3$), 2.07 (s, 3H, $OCOCH_3$), 2.45 (t, J = 2.5 Hz, 1H, C≡CH), 3.62–3.73 (m, 3H, H5 & H6), 3.77 (ddd, J = 11.0 Hz, 7.0 Hz, 3.0 Hz, 1H, $OC\underline{H_2}CH_2$), 3.89 (ddd, J = 10.5 Hz, 9.0 Hz, 8.5 Hz, 1H, H2), 3.96 (ddd, J = 11.5 Hz, 4.5 Hz, 3.0 Hz, 2H, $OC\underline{H'_2}CH_2$), 4.12 (dd, J = 12.0, 2.0, 1H, $OC\underline{H_2}CH'_2$), 4.17 (d, J = 2.5 Hz, 2H, $OCH_2C≡$), 4.24 (dd, J = 11.5, 4.5, 1H, $OC\underline{H'_2}CH'_2$), 4.75 (d, J = 8.5 Hz, 1H, H1), 5.06 (dd, J = 9.5 Hz, 9.5 Hz, 1H, H4), 5.25 (dd, J = 10.5 Hz, 9.5 Hz, 1H, H3), 5.62 (d, J = 9.0 Hz, 1H, NH). **^{13}C NMR** (100 MHz, $CDCl_3$) δ_C = 20.61 ($OCOCH_3$), 20.68 ($OCOCH_3$), 20.74 ($OCOCH_3$), 23.33 ($NHCO\underline{C}H_3$), 54.52 (C2), 58.44 ($O\underline{C}H_2C≡$), 62.10 ($O\underline{C}H_2CH_2$), 68.52 (C4 & $O\underline{C}H_2CH_2$), 69.29 (C6), 71.88 (C5), 72.54 (C3), 74.80 (C≡$\underline{C}$H), 79.51 ($CH_2\underline{C}$≡CH), 100.99 (C1), 169.35 ($NH\underline{C}OCH_3$), 170.30 ($O\underline{C}OCH_3$), 170.70 ($O\underline{C}OCH_3$), 170.85 ($O\underline{C}OCH_3$).

PerAcGlcNAc-2-Alkyne (1-(2-(prop-2-yn-1-yloxy)ethoxy)ethan-1-yl)-2-acetamido-2-deoxy-3,4,6-tri-*O*-acetyl-β-D-glucopyranoside, 45, n = 2)

220 mg glucosamine tetraacetate-derived oxazoline with propargyl diethylene glycol (**38**). Product isolated as a colourless solid (220 mg, 69% yield). Novel compound. **TLC** R_f = 0.20 (19:1 Et_2O:MeOH, $KMnO_4$). **IR** (crystal) υ_{max} = 3297 (m), 3255(m), 3077(w), 2940 (m), 2882 (m), 1736 (s), 1651 (s), 1221 (s), 1085 (s), 1039 (s). **^{1}H NMR** (400 MHz, $CDCl_3$) δ_H = 1.95 (s, 3H, $NHCOC\underline{H_3}$), 2.00 (s, 6H, 2× $OCOCH_3$), 2.08 (s, 3H, $OCOCH_3$), 2.51 (t, J = 2.5 Hz, 1H, C≡$\underline{C}$H), 3.53–3.83 (m, 7H, H5 & 3× $OC\underline{H_2}CH_2$), 3.83–3.90 (m, 2H, $OC\underline{H_2}CH_2$), 4.08–4.17 (m, 2H, H2 & H6), 4.26 (dd, J = 12.0 Hz, 5.0 Hz, 1H, H6′), 4.35 (d, J = 2.5 Hz, 2H, $OCH_2C≡$), 4.81 (d, J = 8.5 Hz, 1H, H1), 4.97–5.14 (m, 2H, H3 & H4), 6.62 (d, J = 9.0 Hz, 1H, NH). **^{13}C NMR** (100 MHz, $CDCl_3$) δ_C = 20.62 ($OCOCH_3$), 20.68 ($OCOCH_3$), 20.77 ($OCOCH_3$), 22.97 ($NHCO\underline{C}H_3$), 53.76 (C2), 58.25 ($O\underline{C}H_2C≡$), 62.22 (C6), 68.44, 69.02, 70.20, 71.79, 72.14, 73.58 (C3 & C4 & C5 & 4× $O\underline{C}H_2CH_2$), 75.33 (C≡$\underline{C}$H), 79.36 ($CH_2\underline{C}$≡CH), 102.17 (C1), 169.26 ($NH\underline{C}OCH_3$), 170.41 ($O\underline{C}OCH_3$), 170.80 ($O\underline{C}OCH_3$), 170.87 ($O\underline{C}OCH_3$). **MS** (ESI) m/z $[M + Na]^+$ Calculated = 496.1789. Found = 496.1776.

PerAcGlcNAc-4-Alkyne (1-(1-*O*-3,6,9,12-Tetraoxapentadec-14-yn-1-yl)-2-acetamido-2-deoxy-3,4,6-tri-*O*-acetyl-β-D-glucopyranoside, 46, n = 4)

220 mg glucosamine tetraacetate-derived oxazoline with propargyl tetraethylene glycol (**39**). Product isolated as a colourless liquid (120 mg, 32% yield). Novel compound. **TLC** R_f = 0.20 (17:3 Et_2O:MeOH, $KMnO_4$). **IR** (oil) υ_{max} = 3276 (m), 2936 (m), 2872(w), 2113 (m), 1743 (s), 1667 (s), 1367 (s), 1224 (s), 1092 (s), 1033 (s). **^{1}H**

NMR (400 MHz, $CDCl_3$) δ_H = 1.95 (s, 3H, NHCOC$\underline{H_3}$), 1.99 (s, 6H, 2× $OCOCH_3$), 2.06 (s, 3H, $OCOCH_3$), 2.43 (t, J = 2.5 Hz, 1H, C≡C$\underline{H}$), 3.54–3.91 (m, 17H, H5 & 8× OC$\underline{H_2}$$CH_2$), 4.02–4.14 (m, 2H, H2 & H6), 4.18 (d, J = 2.5 Hz, 2H, OCH_2C≡), 4.23 (dd, J = 12.5 Hz, 4.5 Hz, 1H, H6′), 4.77 (d, J = 8.5 Hz, 1H, H1), 4.99–5.12 (m, 2H, H3 & H4), 6.66 (d, J = 8.5 Hz, 1H, NH). ^{13}**C NMR** (100 MHz, $CDCl_3$) δ_C = 20.61 ($OCOCH_3$), 20.70 ($OCOCH_3$), 20.75 ($OCOCH_3$), 22.99 (NHCO$\underline{C}H_3$), 53.82 (C2), 58.34 (O$\underline{C}H_2$C≡), 62.18 (C6), 68.68, 68.71, 69.00, 70.15, 70.33, 70.51, 70.55, 70.62, 71.54, 71.63, 73.44 (C3 & C4 & C5 & 8× O$\underline{C}H_2CH_2$), 74.66 (C≡$\underline{C}$H), 79.53 ($CH_2\underline{C}$≡CH), 101.91 (C1), 169.31 (NH$\underline{C}$OCH$_3$), 170.66 (O$\underline{C}$OCH$_3$), 170.72 (O$\underline{C}$OCH$_3$), 170.76 (O$\underline{C}$OCH$_3$). **MS** (ESI) m/z $[M + Na]^+$ Calculated = 584.2314. Found = 584.2303.

7.6.5 Deprotection of Alkynyl Glycosides with Simple Linkers

Acknowledgement: Some of these syntheses and characterisations were carried out with Ian Hazledon.

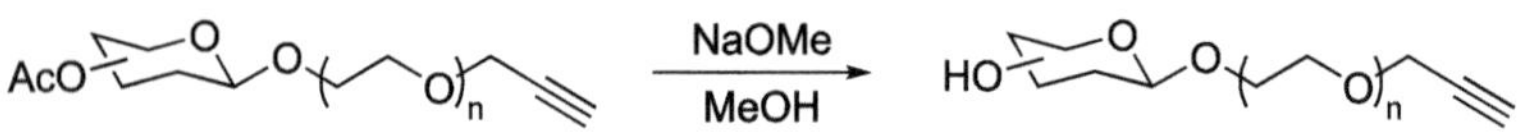

The specified peracetylated alykynyl pyranoside was placed in a round-bottomed flask equipped with a Teflon-coated magnetic stirrer. MeOH (10 mL per 1 g pyranoside) was added and the solution stirred. NaOMe (cat., approximately 25 mg per 1 g pyranoside) was added and the reaction mixture stirred at RT for 2 h. After this time, the solution was neutralised by addition of preconditioned Amberlite IRA120 (H^+) resin until pH < 7 (pH paper). The resin was then filtered and washed with 1:1 MeOH:H_2O (3× 10 mL per 1 g pyranoside). The solvent was removed from the collected filtrate to give the product.

Glc-0-Alkyne (1-*O*-Prop-2-ynyl-β-D-glucopyranoside, 26)

300 mg of 1-*O*-Prop-2-ynyl-2,3,4,6-tetra-*O*-acetyl-β-D-glucopyranoside (**77**). Product a colourless solid in quantitative yield (168 mg). Published compound [22]. **IR** (oil) υ_{max} = 3351 (m), 3277 (m), 2879 (m), 2487 (m), 2119 (w), 1587 (s), 1356 (s), 1073 (s), 1019 (s). ^{1}**H NMR** (400 MHz, D_2O) δ_H = 2.79 (t, J = 2.5 Hz, 1H, C≡CH), 3.17 (dd, J = 9.5 Hz, 8.0 Hz, 1H, H2), 3.26 (dd, J = 10.0 Hz, 9.5 Hz, 1H, H4), 3.31–3.41 (m, 2H, H3 & H5), 3.59 (dd, J = 12.5 Hz, 6.0 Hz, 1H, H6), 3.79 (dd, J = 12.5 Hz, 2.0 Hz, 1H, H6′), 4.32 (dd, J = 16.0 Hz, 3.5 Hz, 1H, OCH_2C≡), 4.37 (dd, J = 16.0 Hz, 3.5 Hz, 1H, OCH'_2 C≡), 4.51 (d, J = 8.0 Hz, 1H, H1). ^{13}**C NMR** (100 MHz, D_2O) δ_C = 56.48 (O$\underline{C}H_2$C≡), 60.62 (C6), 69.47 (C4), 72.80 (C2), 75.62 (C5), 75.90 (C3), 76.33 (C≡$\underline{C}$H), 78.79 ($CH_2\underline{C}$≡CH), 100.46 (C1).

Glc-2-Alkyne (1-*O*-(2-(prop-2-yn-1-yloxy)ethoxy)ethan-1-yl-β-D-glucopyranoside, 81)

90 mg of 1-*O*-(2-(prop-2-yn-1-yloxy)ethoxy)ethan-1-yl-2,3,4,6-tetra-*O*-acetyl-β-D-glucopyranoside (**40**). Product a colourless viscous liquid in quantitative yield (58 mg). Novel compound. **IR** (oil) υ_{max} = 3262 (m), 2878 (m), 2114 (w), 1591 (s), 1352 (s), 1071 (s), 1028 (s). ^{1}**H NMR** (400 MHz, D_2O) δ_H = 2.76 (t, J = 2.5 Hz, 1H, C≡CH), 3.16 (dd, J = 9.5 Hz, 8.0 Hz, 1H, H2), 3.25 (m, 1H, H4), 3.27–3.40 (m, 2H, H3 & H5), 3.51–3.66 (m, 7H, H6 & 3× $OC\underline{H_2}CH_2$), 3.70 (m, 1H, $C1OCH_2$), 3.78 (dd, J = 12.5 Hz, 2.0 Hz, 1H, H6′), 3.92 (dt, J = 11.5 Hz, 4.0 Hz, 1H, $C1OCH_2'$), 4.11 (d, J = 2.5 Hz, 1H, OCH_2C≡), 4.35 (d, J = 8.0 Hz, 1H, H1). ^{13}**C NMR** (100 MHz, D_2O) δ_C = 57.83 ($O\underline{C}H_2C$≡), 60.68 (C6), 68.54 ($O\underline{C}H_2CH_2$), 68.58 ($O\underline{C}H_2CH_2$), 69.34, 69.56, 69.60, 73.04, 75.58, 75.84, 75.94 (C2 & C3 & C4 & C5 & C≡$\underline{C}$H & 2× $O\underline{C}H_2CH_2$), 78.22 ($CH_2\underline{C}$≡CH), 102.16 (C1). **MS** (ESI) m/z $[M + Na]^+$ Calculated = 329.1207. Found =329.1208.

Glc-4-Alkyne (1-*O*-(1-*O*-3,6,9,12-Tetraoxapentadec-14-yn-1-yl)-β-D-glucopyranoside, 82)

80 mg of 1-*O*-3,6,9,12-Tetraoxapentadec-14-yn-1-yl-2,3,4,6-tetra-*O*-acetyl-β-D-glucopyranoside (**41**). Product a colourless liquid in quantitative yield (55 mg). Novel compound. **IR** (oil) υ_{max} = 3363 (m), 3264 (m), 2876 (m), 1590 (s), 1349 (s), 1075 (s), 1040 (s). ^{1}**H NMR** (400 MHz, D_2O) δ_H = 2.76 (t, J = 2.5 Hz, 1H, C≡CH), 3.16 (dd, J = 9.5 Hz, 8.0 Hz, 1H, H2), 3.25 (dd, J = 10.0 Hz, 9.0 Hz, 1H, H4), 3.29–3.39 (m, 2H, H3 & H5), 3.54–3.66 (m, 15H, H6 & 7× $OC\underline{H_2}CH_2$), 3.70 (dt, J = 11.5 Hz, 4.5 Hz, 1H, $C1OCH_2$), 3.78 (dd, J = 12.5 Hz, 2.0 Hz, 1H, H6′), 3.92 (dt, J = 11.5 Hz, 4.0 Hz, 1H, $C1OCH_2'$), 4.11 (d, J = 2.5 Hz, 1H, OCH_2C≡), 4.35 (d, J = 8.0 Hz, 1H, H1). ^{13}**C NMR** (100 MHz, D_2O) δ_C = 57.83 ($O\underline{C}H_2C$≡), 60.69 (C6), 68.58, 68.63, 69.35, 69.45, 69.52, 69.57, 69.62, 73.04, 75.58, 75.85, 75.93 (C2 & C3 & C4 & C5 & C≡$\underline{C}$H & 8× $O\underline{C}H_2CH_2$), 79.25 ($CH_2\underline{C}$≡CH), 102.19 (C1). **MS** (ESI) m/z $[M + Na]^+$ Calculated = 417.1731. Found = 417.1729.

Gal-0-Alkyne (1-*O*-Prop-2-ynyl-β-D-galactopyranoside, 27)

350 mg of 1-*O*-Prop-2-ynyl-2,3,4,6-tetra-*O*-acetyl-β-D-galactopyranoside (**78**). Product a colourless solid in quantitative yield (200 mg). Published compound [21]. **IR** (film) υ_{max} = 3355 (m), 3278 (m), 2878 (m), 2481 (m), 2114 (w), 1586 (s), 1351 (s), 1039 (s). ^{1}**H NMR** (400 MHz, D_2O) δ_H = 2.75 (t, J = 2.5 Hz, 1H, C≡CH), 3.36 (dd, J = 10.0 Hz, 8.0 Hz, 1H, H2), 3.49 (dd, J = 10.0 Hz, 3.5 Hz, 1H, H3), 3.53 (m, 1H, H5), 3.57–3.65 (m, 2H, H6)), 3.76 (d, J = 3.5 Hz, 1H, H4), 4.29 (dd, J = 16.0 Hz, 2.0 Hz, 1H, OCH_2C≡), 4.33 (dd, J = 16.0 Hz, 2.0 Hz, 1H, OCH_2' C≡), 4.40 (d, J = 8.0 Hz, 1H, H1). ^{13}**C NMR** (100 MHz, D_2O) δ_C = 56.40 ($O\underline{C}H_2C$≡), 60.84 (C6), 68.49 (C4), 70.44 (C2), 72.63 (C3), 75.16 (C5), 76.16 (C≡$\underline{C}$H), 78.83 ($CH_2\underline{C}$≡CH), 101.00 (C1). **MS** (ESI) m/z $[M + Na]^+$ Calculated = 241.0683. Found = 241.0685.

Gal-2-Alkyne (1-*O*-(2-(prop-2-yn-1-yloxy)ethoxy)ethan-1-yl-β-D-galactopyranoside, 83)

79 mg of 1-*O*-(2-(prop-2-yn-1-yloxy)ethoxy)ethan-1-yl-2,3,4,6-tetra-*O*-acetyl-β-D-galactopyranoside (**42**). Product a colourless liquid in quantitative yield (51 mg). Novel compound. **IR** (oil) υ_{max} = 3375 (m), 3283 (m), 2877 (m), 2113 (w), 1552 (s), 1066 (s), 1034 (s). ^{1}**H NMR** (400 MHz, D_2O) δ_H = 2.75 (t, *J* = 2.5 Hz, 1H, C≡CH), 3.40 (dd, *J* = 10.0 Hz, 8.0 Hz, 1H, H2), 3.48 – 3.58 (m, 2H, H3 & H5), 3.58–3.67 (m, 8H, H6 & 3× $OC\underline{H_2}CH_2$), 3.70 (m, 1H, $C1OCH_2$), 3.78 (dd, *J* = 3.5 Hz, 1.0 Hz, 1H, H4), 3.94 (dt, *J* = 11.5 Hz, 4.0 Hz, 1H, $C1OCH'_2$), 4.11 (d, *J* = 2.5 Hz, 1H, OCH_2C≡), 4.29 (d, *J* = 8.0 Hz, 1H, H1). ^{13}**C NMR** (100 MHz, D_2O) δ_C = 57.83 ($O\underline{C}H_2C$≡), 60.90 (C6), 68.49, 68.61, 69.35, 69.69(C5 & 4× $O\underline{C}H_2CH_2$), 70.69(C2), 72.62, 75.08 (C3 & C4), 75.94 (C≡$\underline{C}$H), 79.23 ($CH_2\underline{C}$≡CH), 102.76 (C1). **MS** (ESI) m/z $[M + Na]^+$ Calculated = 329.1207. Found = 329.1208.

Gal-4-Alkyne (1-*O*-(1-*O*-3,6,9,12-Tetraoxapentadec-14-yn-1-yl)-β-D-galactopyranoside, 84)

91 mg of 1-*O*-(3,6,9,12-Tetraoxapentadec-14-yn-1-yl-)2,3,4,6-tetra-*O*-acetyl-β-D-galactopyranoside (**43**). Product a colourless liquid in quantitative yield (62 mg). Novel compound. **IR** (oil) υ_{max} = 3384 (m), 3265 (m), 2874 (m), 1071 (s), 1030 (s). ^{1}**H NMR** (400 MHz, D_2O) δ_H = 2.76 (t, *J* = 2.5 Hz, 1H, C≡CH), 3.40 (dd, *J* = 10.0 Hz, 8.0 Hz, 1H, H2), 3.51 (dd, *J* = 10.0 Hz, 3.5 Hz, 1H, H3), 3.53–3.66 (m, 17H, H5 & H6 & 7× $OC\underline{H_2}CH_2$), 3.70 (dt, *J* = 11.5 Hz, 4.5 Hz, 1H, $C1OCH_2$), 3.78 (d, *J* = 3.0 Hz, 1H, H4), 3.94 (dt, *J* = 11.5 Hz, 4.0 Hz, 1H, $C1OCH'_2$), 4.11 (d, *J* = 2.5 Hz, 1H, OCH_2C≡), 4.28 (d, *J* = 8.0 Hz, 1H, H1). ^{13}**C NMR** (100 MHz, D_2O) δ_C = 57.82 ($O\underline{C}H_2C$≡), 60.90 (C6), 68.57, 69.35, 69.45, 69.53, 69.66 (C4 +C≡$\underline{C}$H & 8× $O\underline{C}H_2CH_2$), 75.93 (C3), 75.85 (C2), 75.09 (C5), 75.92 ($CH_2\underline{C}$≡CH), 102.80 (C1). **MS** (ESI) m/z $[M + Na]^+$ Calculated = 417.1731. Found = 417.1735.

GlcNAc-0-Alkyne (1-*O*-Prop-2-ynyl-2-acetamido-2-deoxy-β-D-glucopyranoside, 28)

90 mg of 1-*O*-Prop-2-ynyl-2-acetamido-2-deoxy-3,4,6-tri-*O*-acetyl-β-D-glucopyranoside (**79**). Product a colourless solid in quantitative yield (58 mg). Published compound [22]. **IR** (solid) υ_{max} = 3358 (m), 3265 (m), 2873 (m), 2396 (m), 1633 (s), 1586 (m), 1030 (s). ^{1}**H NMR** (400 MHz, D_2O) δ_H = 1.90 (s, 3H, $NHCOC\underline{H_3}$), 2.77 (t, *J* = 2.5 Hz, 1H, C≡CH), 3.26–3.34 (m, 2H, H4 & H5), 3.42 (dd, *J* = 10.0 Hz, 8.5 Hz, 1H, H3), 3.56 (dd, *J* = 10.0 Hz, 8.5 Hz, 1H, H2), 3.60 (dd, *J* = 12.0 Hz, 5.0 Hz, 1H, H6), 3.78 (dd, *J* = 12.5 Hz, 2.0 Hz, 1H, H6′), 4.27 (d, *J* = 2.5 Hz, 2H, OCH_2C≡), 4.57 (d, *J* = 8.5 Hz, 1H, H1). ^{13}**C NMR** (100 MHz, D_2O) δ_C = 22.13 ($NHCO\underline{C}H_3$), 55.25 (C2), 56.57 ($O\underline{C}H_2C$≡), 60.58 (C6), 69.74 (C4), 73.66 (C3), 75.88 , 76.07 (C5 & C≡$\underline{C}$H), 78.74 ($CH_2\underline{C}$≡CH), 99.32 (C1), 174.65 ($NH\underline{C}OCH_3$). **MS** (ESI) m/z $[M + Na]^+$ Calculated = 282.0948. Found = 282.0949.

GlcNAc-1-Alkyne (1-(2-Prop-2-ynoxy)ethanyl)-2-acetamido-2-deoxy-β-D-glucopyranoside, 85)

50 mg of 1-(2-Prop-2-ynoxy)ethanyl)-2-acetamido-2-deoxy-3,4,6-tri-*O*-acetyl-β-D-glucopyranoside (**80**). Product a yellow solid in quantitative yield (45 mg). Novel compound. **^{1}H NMR** (400 MHz, D_2O) δ_H = 1.89 (s, 3H, $NHCOC\underline{H_3}$), 2.74 (t, J = 2.5 Hz, 1H, C≡CH), 3.27–3.32 (m, 2H, H4 & H5), 3.38 (dd, J = 17.0 Hz, 7.0 Hz, 1H, H3), 3.53 (m, 1H, H2), 3.55–3.62 (m, 3H, H6 & $OC\underline{H_2}CH_2$), 3.62–3.66 (m, 1H, $OC\underline{H_2}CH_2$), 3.77 (d, J = 11.5 Hz, 1H, $OC\underline{H_2}CH_2$), 3.85 (ddd, J = 11.5 Hz, 5.5 Hz, 3.0 Hz, 1H, $OC\underline{H_2}CH_2$), 4.08 (d, J = 3.0 Hz, 2H, OCH_2C≡), 4.40 (d, J = 8.5 Hz, 1H, H1). **^{13}C NMR** (100 MHz, D_2O) δ_C = 22.12 ($NHCO\underline{C}H_3$), 55.42 (C2), 57.85 ($O\underline{C}H_2C$≡), 60.62 ($O\underline{C}H_2CH_2$), 68.62 (C6), 68.73 ($O\underline{C}H_2CH'_2$), 69.79 (C4), 73.71 (C3), 75.75 (C5), 75.86 (C≡$\underline{C}$H), 79.20 ($CH_2\underline{C}$≡CH), 100.99 (C1), 174.56 ($NH\underline{C}OCH_3$).

GlcNAc-2-Alkyne (1-(2-(prop-2-yn-1-yloxy)ethoxy)ethan-1-yl-2-acetamido-2-deoxy-β-D-glucopyranoside, 86)

83 mg of 1-(2-(prop-2-yn-1-yloxy)ethoxy)ethan-1-yl)-2-acetamido-2-deoxy-3,4,6-tri-*O*-acetyl-β-D-glucopyranoside (**45**). Product viscous colourless liquid in quantitative yield (58 mg). Novel compound. **IR** (film) υ_{max} = 3347 (m), 3258 (m), 2871 (m), 2113 (w), 1649 (s), 1562 (m), 1070 (s), 1030(s). **^{1}H NMR** (400 MHz, D_2O) δ_H = 1.91 (s, 3H, $NHCOC\underline{H_3}$), 2.76 (t, J = 2.5 Hz, 1H, C≡CH), 3.28–3.35 (m, 2H, H4 & H5), 3.40 (dd, J = 10.0 Hz, 9.0 Hz, 1H, H3), 3.48–3.68 (m, 9H, H2 & H6 & $C1OCH_2$ & 3× $OC\underline{H_2}CH_2$), 3.79 (dd, J = 12.5 Hz, 1.5 Hz, 1H, H6′), 3.86 (ddd, J = 11.5 Hz, 5.5 Hz, 3.0 Hz, 1H, $C1OCH'_2$), 4.07–4.13 (m, 2H, OCH_2C≡), 4.43 (d, J = 8.5 Hz, 1H, H1). **^{13}C NMR** (100 MHz, D_2O) δ_C = 22.13 ($NHCO\underline{C}H_3$), 55.45 (C2), 57.82 ($O\underline{C}H_2C$≡), 60.66 (C6), 68.63, 68.87, 69.47, 69.58, 69.83, 73.82, 75.81, 75.92(C3 & C4 & C5 & C≡$\underline{C}$H & 4× $O\underline{C}H_2CH_2$), 79.24 ($CH_2\underline{C}$≡CH), 100.96 (C1), 174.49 ($NH\underline{C}OCH_3$). **MS** (ESI) m/z $[M + Na]^+$ Calculated = 370.1472. Found = 370.1467.

GlcNAc-4-Alkyne (1-(1-*O*-3,6,9,12-Tetraoxapentadec-14-yn-1-yl)-2-acetamido-2-deoxy-β-D-glucopyranoside, 87)

69 mg of 1-(1-*O*-3,6,9,12-Tetraoxapentadec-14-yn-1-yl)-2-acetamido-2-deoxy-3,4,6-tri-*O*-acetyl-β-D-glucopyranoside (**46**). Product yellow liquid in quantitative yield (54 mg). Novel compound. **IR** (film) υ_{max} = 3263 (m), 2875 (m), 2113 (w), 1649 (s), 1557 (m), 1351 (m), 1071 (s), 1030(s). **^{1}H NMR** (400 MHz, D_2O) δ_H = 1.90 (s, 3H, $NHCOC\underline{H_3}$), 2.76 (t, J = 2.5 Hz, 1H, C≡CH), 3.27–3.36 (m, 2H, H4 & H5), 3.41 (m, 1H, H3), 3.47–3.69 (m, 17H, H2 & H6 & $C1OCH_2$ & 7× $OC\underline{H_2}CH_2$), 3.79 (dd, J = 12.0 Hz, 1.5 Hz, 1H, H6′), 3.87 (ddd, J = 11.5 Hz, 5.5 Hz, 3.0 Hz, 1H, $C1OCH'_2$), 4.06–4.15 (m, 2H, OCH_2C≡), 4.43 (d, J = 8.5 Hz, 1H, H1). **^{13}C NMR** (100 MHz, D_2O) δ_C = 22.12 ($NHCO\underline{C}H_3$), 55.45 (C2), 57.82 ($O\underline{C}H_2C$≡), 60.66 (C6), 68.58, 68.90, 69.34, 69.49, 69.57, 69.59, 69.63, 69.84, 73.83, 75.81 (C3 & C4 & C5 & 8× $O\underline{C}H_2CH_2$), 75.92 (C≡$\underline{C}$H), 79.23 ($CH_2\underline{C}$≡CH), 100.97 (C1), 174.45 ($NH\underline{C}OCH_3$). **MS** (ESI) m/z $[M + Na]^+$ Calculated = 458.1997. Found = 458.1986.

7.6.6 Synthesis of p-Aminophenyl Glycosides

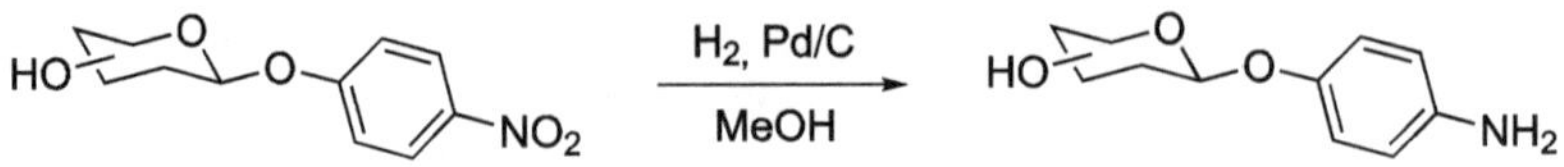

The specified *p*-nitrophenyl glycoside was placed in a round-bottomed flask equipped with a Teflon-coated magnetic stirrer bar. MeOH (10 mL per 100 mg pyranoside) was added and the solution stirred, degassed briefly under reduced pressure and placed under an atmosphere of N_2. 10% Pd on carbon was added (10% of pyranoside weight Pd), and the reaction placed under an atmosphere of H_2 using a balloon. The reaction mixture was stirred for 16 h at room temperature before filtering through Celite and removal of solvent under reduced pressure to give the product.

PAP-Glc (*p*-Aminophenylglucopyranoside, 53)

200 mg of *p*-nitrophenylglucopyranoside (**50**). Product a white solid in quantitative yield (180 mg). Published compound [24]. ^{1}**H NMR** (400 MHz, D_2O) $\delta_H = 3.29$–3.46 (m, 4H, H2 & H3 & H4 & H5), 3.59 (dd, J = 12.5 Hz, 2.0 Hz, 1H, H6), 3.77 (dd, J = 12.5, 2.0 Hz, 1H, H6′), 4.81 (d, J = 7.5 Hz, 1H, H1), 6.67 (d, J = 9.0 Hz, 2H, ArCH), 6.86 (d, J = 9.0 Hz, 2H, ArCH). ^{13}**C NMR** (100 MHz, D_2O) δ_C = 61.21 (C6), 70.10, 73.65, 76.66, 76.69 (C2 & C3 & C4 & C5), 102.34 (C1), 116.36 (ArCH), 117.89 (ArCH), 142.17 (ArC), 150.91 (ArC).

PAP-Gal (*p*-Aminophenylgalactopyranoside, 54)

200 mg of *p*-nitrophenylgalactopyranoside (**51**). Product a white solid in quantitative yield (180 mg). Published compound [24]. ^{1}**H NMR** (400 MHz, D_2O) $\delta_H = 3.55$–3.67 (m, 5H, H2 & H3 & H4 & H5 & H6), 3.80 (d, J = 3.0 Hz, 1H, H6′), 4.75 (d, J = 7.0 Hz, 1H, H1), 6.68 (d, J = 8.5 Hz, 2H, ArCH), 6.80 (d, J = 9.0 Hz, 2H, ArCH). ^{13}**C NMR** (100 MHz, D_2O) δ_C = 60.69, 68.43, 70.55, 72.54, 75.24 (C2 & C3 & C4 & C5 & C6), 101.87 (C1), 117.46 (ArCH), 118.03 (ArCH), 141.50 (ArC), 150.19 (ArC).

PAP-GlcNAc (1-*O*-*p*-Aminophenyl-2-acetamido-2-deoxy-β-D-glucopyranoside, 55)

200 mg of 1-*O*-*p*-nitrophenyl-2-acetamido-2-deoxy-β-D-glucopyranoside (**52**). Product a white solid in quantitative yield (182 mg). Published compound [25]. ^{1}**H NMR** (400 MHz, D_2O) δ_H = 1.89 (s, 3H, NHCO$\underline{CH_3}$), 3.36–3.51 (m, 3H, H3 & H4 & H5), 3.78 (m, 1H, H6), 3.75–3.83 (m, 2H, H2 & H6′), 4.83 (d, J = 8.5 Hz, 1H, H1), 6.64 (d, J = 9.0 Hz, 2H, ArCH), 6.78 (d, J = 9.0 Hz, 2H, ArCH). ^{13}**C NMR** (100 MHz, D_2O) δ_C = 22.05 (NHCO$\underline{C}H_3$), 55.50 (C2), 60.48 (C6), 69.63, 73.53, 75.97 (C3 & C4 & C5), 100.73 (C1), 117.41 (ArCH), 118.22 (ArCH), 141.79 (ArC), 150.16 (ArC), 174.70 (NH$\underline{C}OCH_3$).

7.6.7 *Synthesis of Alkynyl Linker Carboxylic Acids*

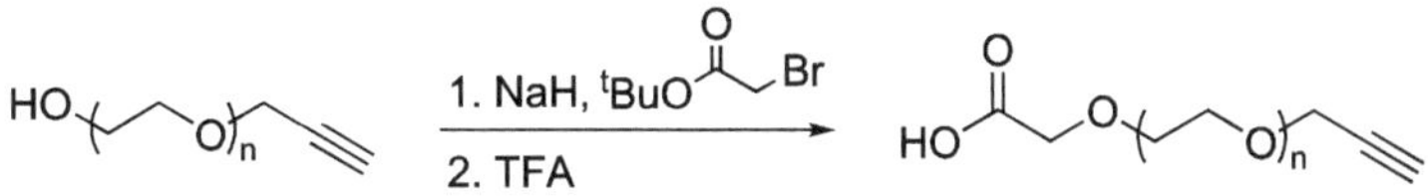

The specified alkynyl alcohol (1 equivs.) was placed in a three-necked round-bottomed flask equipped with a Teflon-coated magnetic stirrer. Anhydr. THF (20 mL per 1 g alcohol) was added, and the solution stirred and cooled to 0°C with an ice bath. NaH (60% suspension in mineral oil, 2 equivs.) was added and the mixture stirred for 10 min. After this, *t*-butylbromoacetate (2 equivs.) was added, the ice bath removed and stirring continued at room temperature for 16 h. After this time, the reaction mixture was carefully diluted with H_2O (equal volume to THF), the organic layer separated and the aq. extracted with Et_2O (3× equal volume to THF). The combined organics were dried over Na_2SO_4, filtered and the solvent removed under reduced pressure. The resultant residue was purified by FCC (3:1 hexane:EtOAc) to give the *t*-butyl ester of the product. This was taken up in CH_2Cl_2 (5 mL per 1 g alcohol) in a round-bottomed flask equipped with a Teflon-coated magnetic stirrer and hydrolysed by addition of an equal volume of TFA. After stirring for 2 h at room temperature, reduction under reduced pressure yielded the product.

Acid-2-Alkyne (2-(2-(2-(2-propyn-1-yloxy)ethoxy)ethoxy)-acetic Acid, 88), n = 2

250 mg of 2-(2-(prop-2-yn-1-yloxy)ethoxy)ethan-1-ol (**38**). Product a clear liquid (196 mg, 56% yield). Published compound [26]. ^{1}**H NMR** (400 MHz, $CDCl_3$) δ_H = 2.45 (t, J = 2.5 Hz, 1H, C≡CH), 3.70–3.79 (m, 6H, 4× CH_2O), 4.21 (d, J = 2.5 Hz, 2H, OCH_2C≡). ^{13}**C NMR** (100 MHz, $CDCl_3$) δ_C = 58.39 (O$\underline{C}H_2$C≡), 68.52 (C(O)CH_2), 68.84, 70.21, 70.40, 71.27 (4× CH_2O), 74.85 (C≡$\underline{C}$H), 79.20 ($CH_2\underline{C}$≡CH), 173.88 (CO_2H).

Acid-4-Alkyne (3,6,9,12,15-Pentaoxaoctadec-17-ynoic acid, 57), n = 4

500 mg of 3,6,9,12-Tetraoxapentadec-14-yn-1-ol (**39**). Product a clear liquid (224 mg, 36% yield). Novel compound. ^{1}**H NMR** (400 MHz, $CDCl_3$) δ_H = 2.43 (t, J = 2.5 Hz, 1H, C≡CH), 3.63–3.78 (m, 16H, 8× CH_2O), 4.18 (s, 2H, C(O)CH_2), 4.20 (d, J = 2.5 Hz, 2H, OCH_2C≡). ^{13}**C NMR** (100 MHz, $CDCl_3$) δ_C = 58.32 (O$\underline{C}H_2$C≡), 68.82 (C(O)CH_2), 70.22, 70.25, 70.47, 71.27 (8× CH_2O), 74.79 (C≡$\underline{C}$H), 79.22 ($CH_2\underline{C}$≡CH), 173.38 (CO_2H).

7.6.8 Synthesis of Alkynyl Glycosides with Aromatic Linkers

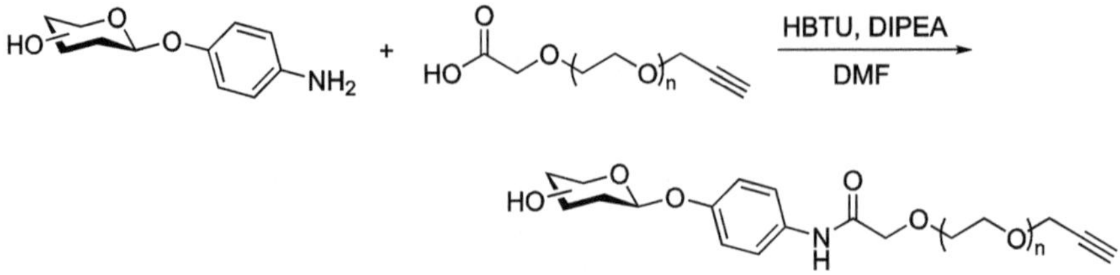

The specified alkynyl polyether carboxylic acid (2 equivs.) was placed in a three-necked round-bottomed flask equipped with a Teflon-coated magnetic stirrer. Anhydr. DMF (1 mL per 100 mg glycoside) was added, the solution stirred and HBTU (1.18 equivs.) added followed by DIEA (2 equivs.) and the mixture stirred for 15 min at room temperature. After this, the specified *p*-aminophenyl glycoside (1 equivs.) was added as a solution in anhydr. DMF (1 mL per 100 mg glycoside) and the reaction mixture stirred for 16 h at RT. After this, the solvent was removed under reduced pressure and the mixture purified by FCC (4:1 CH_2Cl_2:MeOH), and fractions containing the product were reduced under vacuum and further purified by preparative HPLC (5-60% solvent B) to give the product. Only distinguishable peaks in the NMR spectra have been assigned.

Gal-Ar-2-Alkyne (89)

200 mg of *p*-aminophenylgalactopyranoside (**54**) with Acid-2-Alkyne (**88**). Product a waxy solid (194 mg, 58% yield). Novel compound. **TLC** R_f = 0.40 (4:1 CH_2Cl_2:MeOH, $KMnO_4$). **^{1}H NMR** (400 MHz, CD_3OD) δ_H = 2.83 (t, *J* = 2.5 Hz, 1H, C≡CH), 3.57 (dd, *J* = 9.5 Hz, 3.5 Hz, 1H), 3.64–3.81 (m, 12H), 3.90 (m, 1H), 4.11 (s, 1H, C=OCH_2O), 4.14 (d, *J* = 2.5 Hz, 2H, OCH_2C≡), 4.82 (d, *J* = 7.5 Hz, 1H, H1), 7.10 (d, *J* = 9.0 Hz, 2H, ArCH), 7.51 (d, *J* = 9.0 Hz, 2H, ArCH). **^{13}C NMR** (100 MHz, CD_3OD) δ_C = 57.60 ($O\underline{C}H_2C$≡), 60.99, 68.54, 68.80, 69.89, 69.98, 70.02, 70.63, 70.86, 73.43, 74.61, 75.54, 101.86 (C1), 116.70 (ArCH), 121.82 (ArCH), 131.89 (ArC), 154.80 (ArC), 169.40 (C=O).

GlcNAc-Ar-2-Alkyne (90)

200 mg of 1-*O*-*p*-aminophenyl-2-acetamido-2-deoxy-β-D-glucopyranoside (**55**) with Acid-2-Alkyne (**88**). Product a waxy solid (218 mg, 68% yield). Novel compound. **TLC** R_f = 0.40 (4:1 CH_2Cl_2:MeOH, $KMnO_4$). **^{1}H NMR** (400 MHz, D_2O) δ_H = 1.93 (s, 3H, CH_3), 2.74 (m, *J* = 2.5 Hz, 1H, C≡CH), 3.42–3.76 (m, 12H), 3.81 – 3.91 (m, 2H), 4.03 (s, 2H, OCH_2C≡), 4.13 (s, 2H, C=OCH_2O), 5.05 (d, *J* = 8.5 Hz, 1H, H1), 7.00 (d, *J* = 8.0 Hz, 2H, ArCH), 7.29 (d, *J* = 8.0 Hz, 2H, ArCH).

Glc-Ar-4-Alkyne (58)

200 mg of *p*-aminophenylglucopyranoside (**53**) with Acid-4-Alkyne (**57**). Product a viscous liquid (130 mg, 32% yield). Novel compound. **TLC** R_f = 0.40 (4:1 CH_2Cl_2:MeOH, $KMnO_4$). **^{1}H NMR** (400 MHz, D_2O) δ_H = 2.77 (s, 1H, C≡CH),

3.34–3.60 (m, 14H), 3.61–3.70 (m, 5H), 3.71–3.76 (m, 2H), 3.83 (d, J = 12.5 Hz, 1H), 4.07 (s, 1H, C=OCH_2O), 4.13 (d, J = 2.5 Hz, 2H, $OCH_2C\equiv$), 5.01 (d, J = 7.5 Hz, 1H, H1), 7.07 (d, J = 9.0 Hz, 1H, ArCH), 7.32 (d, J = 9.0 Hz, 2H, ArCH).

Gal-Ar-4-Alkyne (59)

200 mg of *p*-aminophenylgalactopyranoside (**54**) with Acid-4-Alkyne (**57**). Product a viscous liquid (160 mg, 39% yield). Novel compound. **TLC** R_f = 0.40 (4:1 CH_2Cl_2:MeOH, $KMnO_4$). ^{1}**H NMR** (400 MHz, CD_3OD) δ_H = 2.85 (t, J = 2.5 Hz, 1H, C≡CH), 3.56–3.82 (m, 21H), 3.92 (d, J = 3.0 Hz, 1H), 4.11–4.17 (m, 4H, C=OCH_2O & $OCH_2C\equiv$), 4.84 (d, J = 8.0 Hz, 1H, H1), 7.51 (d, J = 9.0 Hz, 2H, ArCH), 7.10 (d, J = 9.0 Hz, 2H, ArCH). ^{13}**C NMR** (100 MHz, CD_3OD) δ_C = 54.39, 57.63, 60.96, 68.55, 68.76, 69.65, 69.74, 69.76, 69.84, 69.86, 69.92, 70.44, 70.86, 73.38, 74.78, 75.47, 79.12, 101.79(C1), 116.75 (ArCH), 121.81 (ArCH), 131.96 (ArC), 154.71 (ArC), 169.43 (C=O).

GlcNAc-Ar-4-Alkyne (60)

200 mg of 1-*O*-*p*-aminophenyl-2-acetamido-2-deoxy-β-D-glucopyranoside (**55**) with Acid-4-Alkyne (**57**). Product a waxy solid (90 mg, 24% yield). Novel compound. **TLC** R_f = 0.40 (4:1 CH_2Cl_2:MeOH, $KMnO_4$). ^{1}**H NMR** (400 MHz, D_2O) δ_H = 1.94 (s, 3H, CH_3), 2.77 (m, J = 2.5 Hz, 1H, C≡CH), 3.43–3.76 (m, 19H), 3.81 – 3.91 (m, 2H), 4.07 (d, J = 2.5 Hz, 2H, $OCH_2C\equiv$), 4.13 (s, 2H, C=OCH_2O), 5.05 (d, J = 8.5 Hz, 1H, H1), 7.01 (d, J = 9.0 Hz, 2H, ArCH), 7.31 (d, J = 9.0 Hz, 2H, ArCH).

7.7 Cellular Studies

Cell Culture

Acknowledgement: the studies with OPCs were carried out with Dr Bangfu Zhu (University of Cardiff).

Trypsinised cells were counted using a Trypan blue exclusion method. Gels were made as previously outlined (Sect. 7.5.3) in 96-well tissue culture plates. The gelled discs were seeded with 10 000 cells and supplemented with 100 μL supplemented media. Murine fibroblasts (NIH 3T3, *ATCC*, UK) were cultured in supplemented high glucose Dulbecco's Modified Eagle Medium with 10% v/v foetal bovine serum, 1% v/v penicillin-streptomycin, 2.5% v/v L-glutamine and 2.5% v/v 1 M HEPES buffer. Murine OPCs were isolated and cultured according to literature procedures [27]. All cells were maintained at 37 °C, 5% CO_2 and 100% relative humidity with the media being changed on every third day. Cells were passaged using trypsin, with cells at passage 4 being used for this study.

7.7.1 Live/Dead Assay

For live cell imaging, aq. calcein-AM (5 μL) was added to the sample, followed by aq. propidium iodide (5 μL) and the sample imaged using a *Leica* DM IRBE inverted epifluorescence microscope. For fixation prior to imaging, samples were washed with 50 μL PBS buffer before incubation with 40 μL 1% aq. glutaraldehyde for 30 min at 37°C before storage under 100 μL PBS buffer at 4°C.

Green fluorescence is calcein, staining live cells. Red fluorescence is propridium iodide, staining dead cells.

No red fluorescence was observed for samples decorated with Glc-0-alkyne, Gal-0-alkyne, or GlcNAc-0-alkyne.

hSAF-p1 & hSAF-p2 live

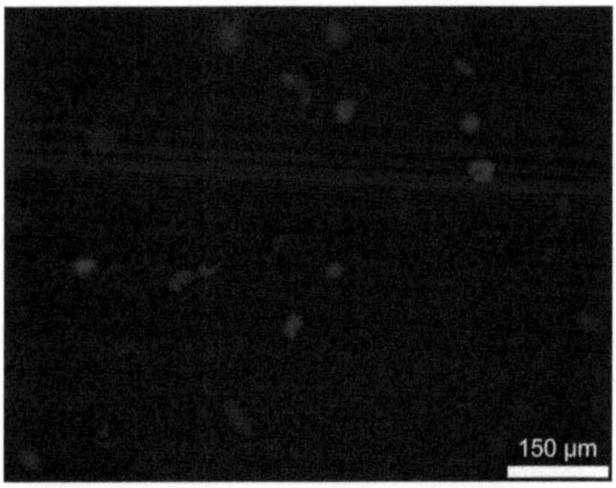

hSAF-p1 & hSAF-p2 dead

hSAF-p1-K1Z & hSAF-p2 live

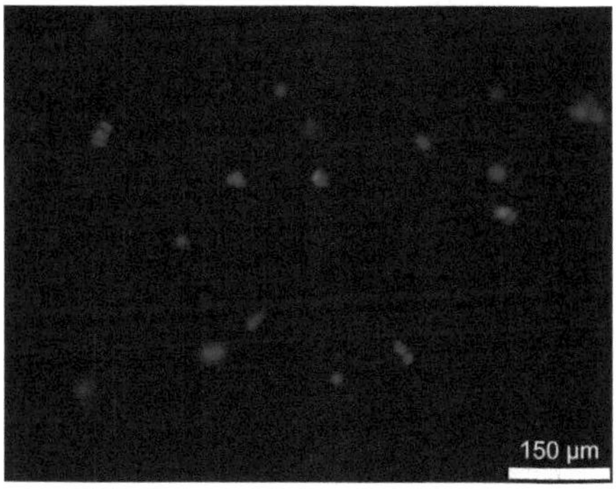

hSAF-p1-K1Z & hSAF-p2 dead

hSAF-p1-K1Z+Glc-0 & hSAF-p2 live

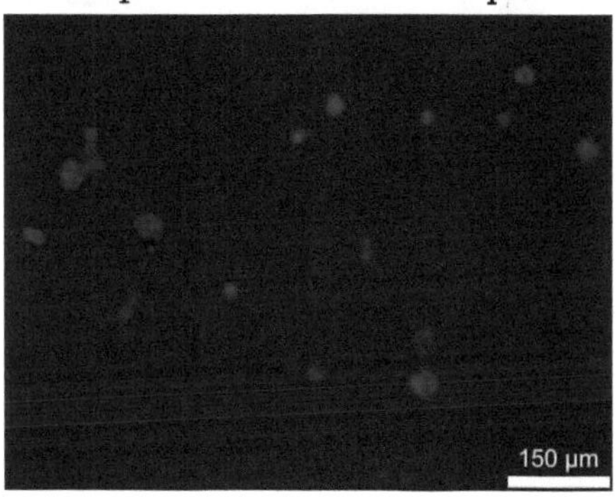

hSAF-p1-K1Z+Gal-0 & hSAF-p2 live

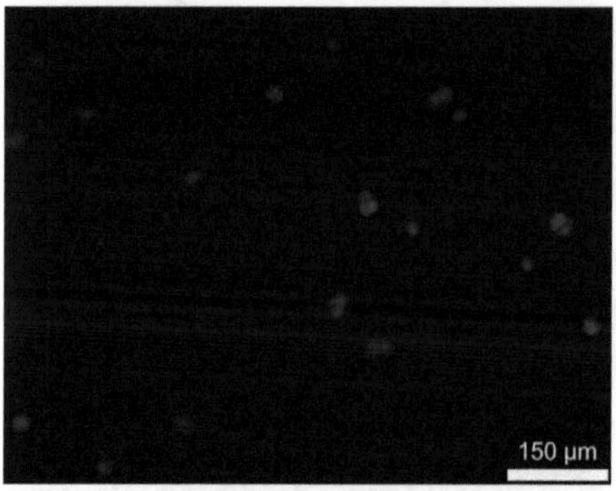

hSAF-p1-K1Z+GlcNAc-0 & hSAF-p2 live

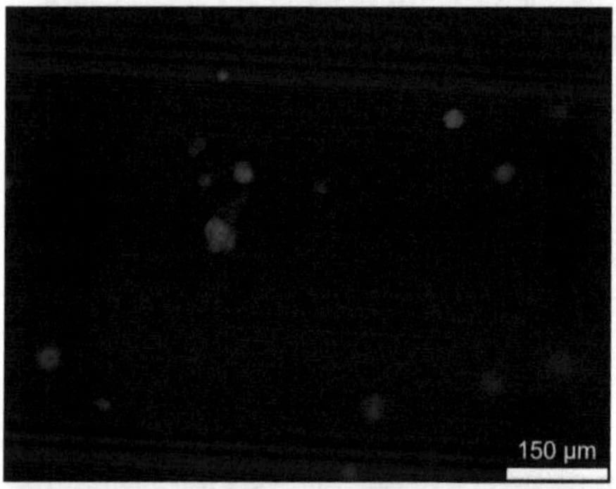

BME live

BME dead

TCP live

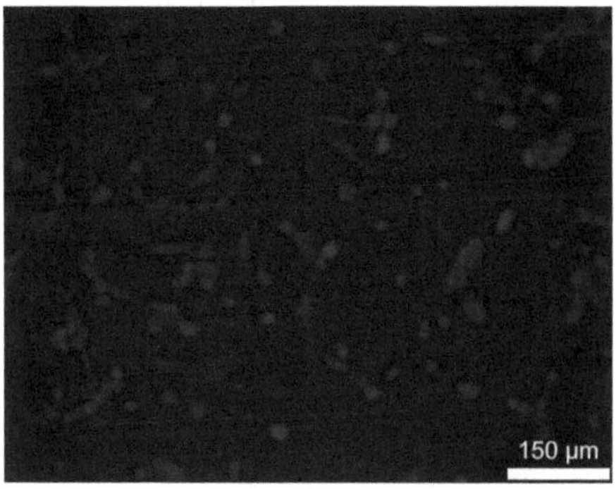

TCP dead

7.7.2 *Metabolic Assay*

The metabolic assay was carried out at days 0, 3, 7 (and 14 for studies with fibroblasts) of cell culture, in triplicate. 10 μL of 250 mg mL^{-1} aq. MTT was added to the samples, which were then incubated at 37 °C. The following day excess media in the sample was removed to leave 25 μL before addition of 100 μL of dimethylsulfoxide (DMSO), mixing and incubation at 37 °C for 30 min, repeated DMSO addition and incubation and transfer of 150 μL of the solution to measure absorbance at 560 nm in the visible spectrum.

7.8 Enzymatic Carbohydrate Couplings

NaPi buffer was prepared as a 10× solution to give a final concentration of 100 mM sodium phosphate in H_2O. HEPES buffer was prepared as a 5× solution to give a final concentration of 20 mM HEPES in H_2O. Tris buffer was prepared as a 10× solution to give a final concentration of 50 mM Tris in H_2O. Sep-Paks were C18 Vac RC/500 mg model from *Waters* and used under gravity flow.

7.8.1 Coupling of Glucose

Synthesis of CelB-Ar-4-alkyne (62) and CelT-Ar-4-alkyne (63)

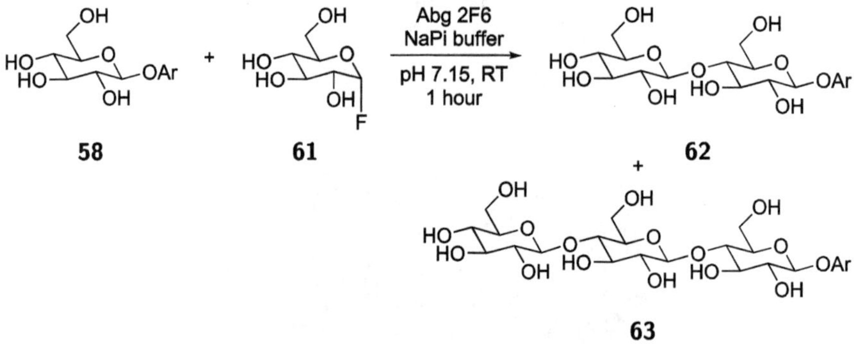

Adapted from a literature procedure [28]. Enzyme Abg 2F6 [28] (1 mg) was dissolved in NaPi buffer (1 mL) at pH 7.15, followed by Glc-Ar-4-alkyne (**58**, 21.6 mg, 40 μmol) and GlcF (**61**, 9.2 mg, 60 μmol). The reaction mixture was left at RT for 1 h. After this time, precipitate was removed by centrifugation and the supernatant placed on a C18 Sep-Pak, which was washed with H_2O (5 mL) and then the products eluted with 3 mL each of 10%, 20%, 30%, 40%, 50%, and 60% MeOH in H_2O. The UV-active (by TLC) fractions—20% and 30%—were collected and reduced under vacuum to give a viscous liquid. This was purified by semi-preparative HPLC (15-25% MeCN in H_2O) to give 10 mg of CelB-Ar-4-alkyne (**62**, 14 μmol, 35% yield) and 3.5 mg of CelT-Ar-4-alkyne (**63**, 4 μmol, 10% yield). Novel compounds. **CelB**: [1]**H NMR** (400 MHz, D_2O) $\delta_H = 2.75$ (s, 1H, C≡CH), 3.20 (dd, $J = 8.5$ Hz, 8.5 Hz, 1H), 3.26 – 3.77 (m, 25H), 3.81 (d, $J = 12.0$ Hz, 1H), 3.86 (d, $J = 12.0$ Hz, 1H), 4.04 (s, 2H, C=OCH$_2$O), 4.10 (s, 2H, OCH$_2$C≡), 4.41 (d, $J = 8.0$ Hz, 1H, Glc2-H1),

5.00 (d, J = 8.0 Hz, 1H, Glc1-H1), 7.04 (d, J = 8.5 Hz, 1H, ArCH)), 7.30 (d, J = 8.5 Hz, 2H, ArCH). **MS** (MALDI) m/z $[M + Na]^+$ Calculated = 728.274. Found = 728.308. **CelT**: ^{1}**H NMR** (400 MHz, D_2O) δ_H = 2.74 (s, 1H, C≡CH), 3.14–3.41 (m, 5H), 3.45–3.82 (m, 27H), 3.87 (d, J = 12.0 Hz, 2H), 4.04 (s, 2H, C=OCH_2O), 4.11 (s, 2H, OCH_2C≡), 4.39 (d, J = 8.0 Hz, 1H, Glc3-H1), 4.44 (d, J = 8.0 Hz, 1H, Glc2-H1), 5.01 (d, J = 8.0 Hz, 1H, Glc1-H1),7.04 (d, J = 9.0 Hz, 1H, ArCH)), 7.29 (d, J = 9.0 Hz, 2H, ArCH). **MS** (MALDI) m/z $[M + Na]^+$ Calculated = 890.326. Found = 890.330.

7.8.2 *Coupling of Galactose*

Synthesis of Lac-Ar-4-alkyne (65)

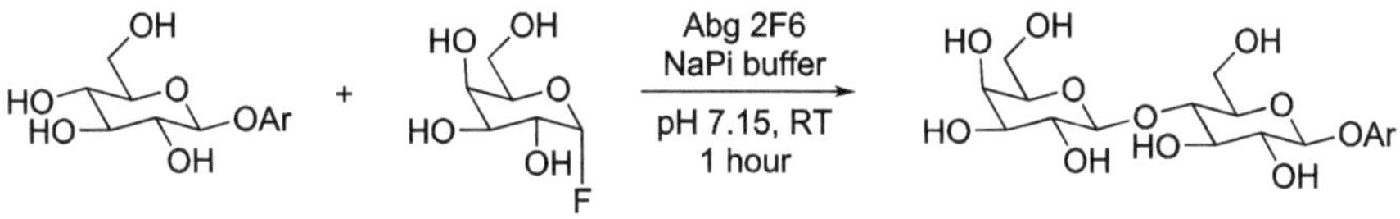

58 **64** **65**

Adapted from a literature procedure [28]. Enzyme Abg 2F6 [28] (1 mg) was dissolved in NaPi buffer (1 mL) at pH 7.15, followed by Glc-Ar-4-alkyne (**58**, 5.4 mg, 10 μmol) and GalF (**64**, 2.7 mg, 15 μmol). The reaction mixture was left at RT for 1 h. After this time, precipitate was removed by centrifugation and the supernatant placed on a C18 Sep-Pak, which was washed with H_2O (5 mL) and then the products eluted with 3 mL each of 10%, 20%, 30%, 40%, and 50% MeOH in H_2O. The UV-active (by TLC) fraction—30%—was reduced under vacuum to give a viscous liquid. This was purified by semi-preparative HPLC (15-25% MeCN in H_2O) to give 3.8 mg of Lac-Ar-4-alkyne (**65**, 5.2 μmol, 52% yield). Novel compound. ^{1}**H NMR** (400 MHz, D_2O) δ_H = 2.74 (s, 1H, C≡CH), 3.38–3.76 (m, 26H), 3.81 (d, J = 3.0 Hz, 1H), 3.87 (d, J = 12.0 Hz, 1H), 4.04 (s, 2H, C=OCH_2O), 4.11 (s, 2H, OCH_2C≡), 4.36 (d, J = 8.0 Hz, 1H, Gal-H1), 5.02 (d, J = 8.0 Hz, 1H, Glc-H1), 7.05 (d, J = 8.5 Hz, 1H, ArCH)), 7.30 (d, J = 8.5 Hz, 2H, ArCH). **MS** (MALDI) m/z $[M + Na]^+$ Calculated = 728.274. Found = 728.313.

Synthesis of diGal-Ar-4-alkyne (67)

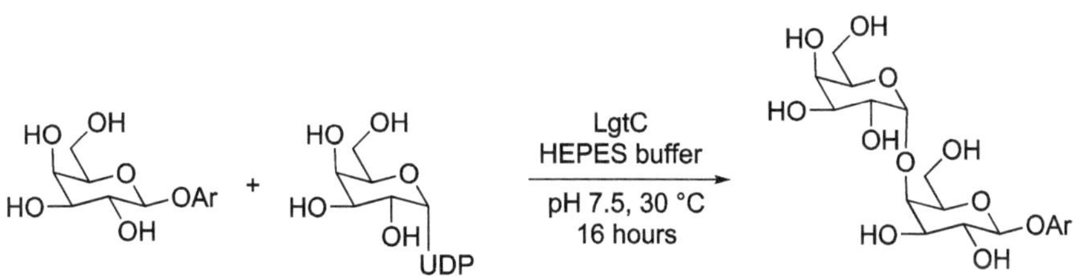

59 **66** **67**

Adapted from a literature procedure [29]. Enzyme Lgtc [30] (20 μL of solution at unknown concentration) was added to HEPES buffer (1 mL) at pH 7.5 with 50 mM KCl, 5 mM $MnCl_2$, and 5 mM DTT, followed by Gal-Ar-4-alkyne (**59**, 10.8 mg, 20 μmol) and UDP-Gal (**66**, 18.1 mg, 30 μmol). The reaction mixture was warmed to 30°C and left for 16 h. After this time, precipitate was removed by centrifugation and the supernatant placed on a C18 Sep-Pak, which was washed with H_2O (5 mL) and then the products eluted with 3 mL each of 10%, 20%, 30%, 40%, and 50% MeOH in H_2O. The UV-active (by TLC) fraction—30%—was reduced under vacuum to give a viscous liquid. This was purified by semi-preparative HPLC (15-25% MeCN in H_2O) to give 6.8 mg of diGal-Ar-4-alkyne (**67**, 9.6 μmol, 48% yield). Novel compound. **^{1}H NMR** (400 MHz, D_2O) δ_H = 2.74 (t, J = 2.5 Hz, C≡CH), 3.44 – 3.52 (m, 6H), 3.52–3.57 (m, 4H), 3.57–3.62 (m, 4H), 3.62–3.67 (m, 2H), 3.67–3.77 (m, 7H) 3.77–3.84 (m, 2H), 3.92 (d, J = 3.0 Hz, 1H), 3.99 (s, 1H), 4.04 (d, J = 2.5 Hz, 2H, OCH_2C≡), 4.10 (s, 2H, $C{=}OCH_2O$), 4.27 (t, J = 6.5 Hz, 1H), 4.87 (d, J = 4.0 Hz, 1H, Gal1-H1), 4.95 – 5.05 (m, Gal2-H1), 7.05 (d, J = 9.0 Hz, 1H, ArCH)), 7.29 (d, J = 9.0 Hz, 2H, ArCH). **MS** (MALDI) m/z $[M + Na]^+$ Calculated = 728.274. Found = 728.417.

Synthesis of LacNAc-Ar-4-alkyne (68)

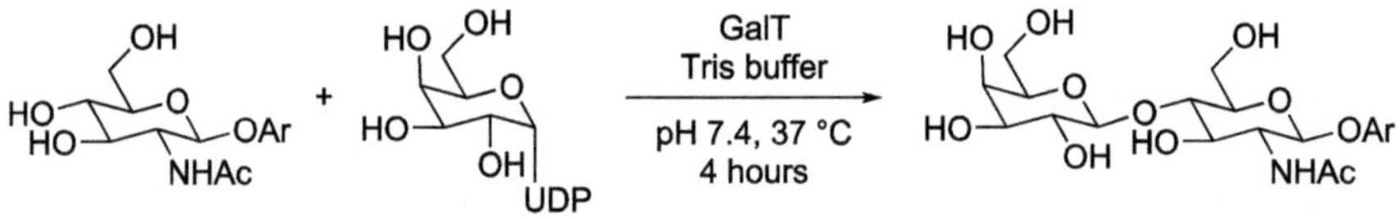

60 **66** **68**

β-1,4-Galactosyltransferase (10 μL of solution at 10 UmL^{-1}, so 10 mU) was added to Tris buffer (1 mL) at pH 7.5 with 20 mm $MnCl_2$, followed by BSA (2 mg), alkaline phosphatase (2 μL of 10 $U\mu L^{-1}$ solution, so 20 U), GlcNAc-Ar-4-alkyne (**60**, 17.4 mg, 30 μmol), and UDP-Gal (**66**, 21.9 mg, 36 μmol). The reaction mixture was warmed to 37°C and left for 16 h. After this time, precipitate was removed by centrifugation and the supernatant placed on a C18 Sep-Pak, which was washed with H_2O (10 mL) and then the products eluted with 5 mL each of 10%, 20%, 30%, 40%, and 50% MeOH in H_2O. The UV-active (by TLC) fractions—20% and 30%—were combined and reduced under vacuum to give a viscous liquid. This was purified by preparative HPLC (15-25% MeCN in H_2O) to give 16.0 mg of LacNAc-Ar-4-alkyne (**68**, 21 μmol, 72% yield). Novel compound. **^{1}H NMR** (400 MHz, D_2O) δ_H = 1.90 (s, 3H, $C{=}OCH_3$), 2.74 (t, J = 2.0 Hz, 1H, C≡CH), 3.38–3.78 (m, 25H), 3.80 (d, J = 3.0 Hz, 1H), 3.84–3.93 (m, 2H), 4.03 (d, J = 2.0 Hz, 2H, OCH_2C≡), 4.10 (s, 2H, $C{=}OCH_2O$), 4.38 (d, J = 8.0 Hz, 1H, Gal-H1), 5.04 (d, J = 8.5 Hz, 1H, GlcNAc-H1), 6.98 (d, J = 9.0 Hz, 1H, ArCH)), 7.28 (d, J = 9.0 Hz, 2H, ArCH). **MS** (MALDI) m/z $[M + Na]^+$ Calculated = 769.300. Found = 769.437.

7.8.3 Coupling of N-Acetylneuramic Acid

Synthesis of SiaGal-Ar-4-alkyne (67)

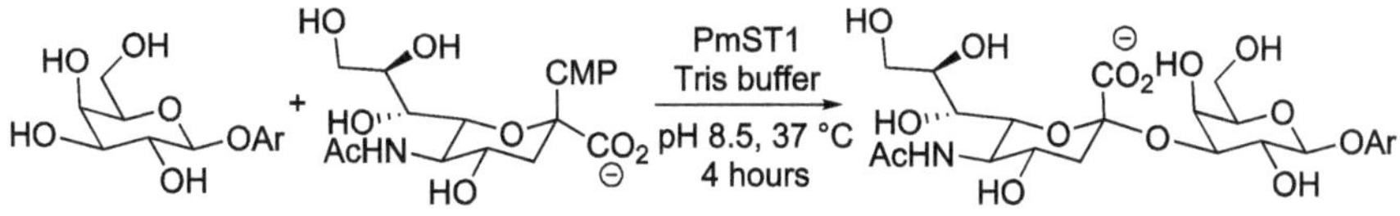

59 **69** **70**

Adapted from a literature procedure [31]. Enzyme PmST1 M144D [31] (15.6 μL of solution at 3.2 mgmL^{-1}, so 50 μg) was added to Tris buffer (1 mL) at pH 8.5 with 10 mM $MgCl_2$, followed by Gal-Ar-4-alkyne (**59**, 5.4 mg, 10 μmol) and cytidine-5'-monophospho-*N*-acetylneuraminic acid (CMP-NeuNAc) (**69**, 9 mg, 15 μmol). The reaction mixture was warmed to 37°C and left for 4 h. After this time, precipitate was removed by centrifugation and the supernatant placed on a C18 Sep-Pak, which was washed with H_2O (5 mL) and then the products eluted with 3 mL each of 10%, 20%, 30%, 40%, and 50% MeOH in H_2O. The UV-active (by TLC) fractions—20% and 30%—were reduced under vacuum to give a viscous liquid. This was purified by semi-preparative HPLC (15-25% MeCN in H_2O) to give 2.9 mg of SiaGal-Ar-4-alkyne (**70**, 3.5 μmol, 35% yield). Novel compound. ^{1}**H NMR** (400 MHz, D_2O) δ_H = 1.90 (s, 3H, C=OCH$_3$), 1.92 (s, 1H, Sia-H3), 2.65 (dd, J = 12.5 Hz, 4.5 Hz, 1H, Sia-H3′), 2.74 (t, J = 2.0 Hz, 1H, C≡CH), 3.41–3.78 (m, 26H), 3.88 – 3.93 (m, 1H), 4.04 (d, J = 2.0 Hz, 2H, OCH$_2$C≡), 4.10 (s, 3H), 5.01 (d, J = 8.0 Hz, 1H, Gal-H1), 7.04 (d, J = 9.0 Hz, 1H, ArCH)), 7.28 (d, J = 9.0 Hz, 2H, ArCH). **MS** (MALDI) m/z [M-H]$^-$ Calculated = 833.320. Found = 833.578.

Synthesis of SiaLacNAc-Ar-4-alkyne (71)

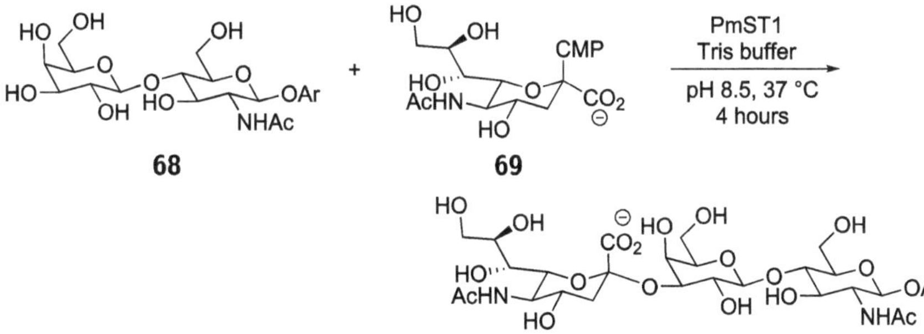

68 **69** **71**

Adapted from a literature procedure [31]. Enzyme PmST1 M144D [31] (10.5 μL of solution at 3.2 mgmL^{-1}, so 30 μg) was added to Tris buffer (1 mL) at pH 8.5 with 10 mM $MgCl_2$, followed by LacNAc-Ar-4-alkyne (**68**, 5 mg, 6 μmol) and CMP-NeuNAc (**69**, 6 mg, 9 μmol). The reaction mixture was warmed to 37°C and left for 4 hours. After this time, precipitate was removed by centrifugation and the supernatant placed on a C18 Sep-Pak, which was washed with H_2O (5 mL) and then the products eluted with 3 mL each of 10%, 20%, 30%, 40%, and 50% MeOH in H_2O. The UV-active (by TLC) fractions—20–40%—were reduced under vacuum to give a viscous liquid. This was purified by semi-preparative HPLC (15-25% MeCN in H_2O) to give 3.6 mg of SiaGal-Ar-4-alkyne (**70**, 3.5 μmol, 58% yield). Novel compound. **^{1}H NMR** (400 MHz, D_2O) δ_H = 1.90 (s, 3H, C=OCH$_3$), 1.92 (s, 1H, Sia-H3), 2.63 (dd, J = 12.5 Hz, 4.5 Hz, 1H, Sia-H3′), 2.73 (t, J = 2.0 Hz, 1H, C≡CH), 3.41–3.79 (m, 29H), 3.81 – 3.93 (m, 4H), 4.03 (m, 3H), 4.10 (s, 2H, C=OCH$_2$O), 4.45 (d, J = 8.0 Hz, 1H, Gal-H1), 5.03 (d, J = 8.5 Hz, 1H, GlcNAc-H1), 6.97 (d, J = 9.0 Hz, 1H, ArCH)), 7.27 (d, J = 9.0 Hz, 2H, ArCH). **MS** (MALDI) m/z [M-H]$^-$ Calculated = 1036.399. Found = 1036.70.

7.8.4 Coupling of Fucose

Synthesis of Le^X-Ar-4-alkyne (73)

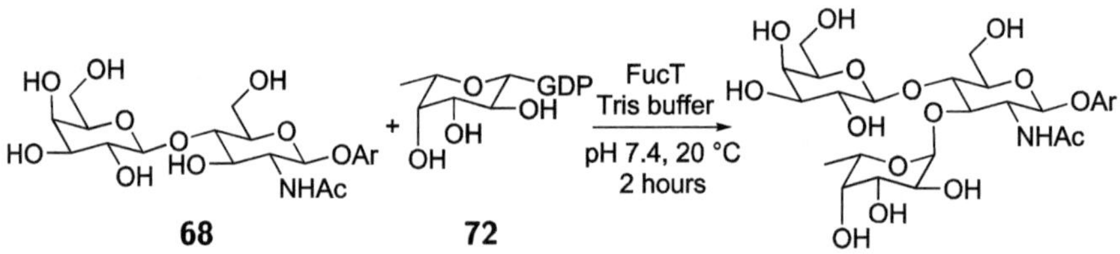

68 **72** **73**

Adapted from a literature procedure [32]. Expressed and purified enzyme FucT (8 μL of solution at 6.3 mgmL^{-1}, so 50 μg) was added to Tris buffer (1 mL) at pH 7.4 with 20 mm MnCl$_2$, followed by BSA (1 mg), alkaline phosphatase (2 μL of 10 UμL^{-1} solution, so 20 U), and LacNAc-Ar-4-alkyne (**68**, 5 mg, 6.7 μmol) and GDP-Fuc (**72**, 5 mg, 8 μmol). The reaction was left for 2 h at RT. After this time, precipitate was removed by centrifugation and the supernatant placed on a C18 Sep-Pak, which was washed with H_2O (5 mL) and then the products eluted with 3 mL each of 10%, 20%, 30%, 40%, and 50% MeOH in H_2O. The UV-active (by TLC) fractions—30% and 40%—were reduced under vacuum to give a viscous liquid. This was purified by semi-preparative HPLC (15-25% MeCN in H_2O) to give 2.3 mg of Le^X-Ar-4-alkyne (**73**, 2.6 μmol, 43% yield). Novel compound. **^{1}H NMR** (400 MHz, D_2O) δ_H = 1.06 (d, J = 6.5 Hz, 3H, Fuc-CH$_3$), 1.89 (s, 3H, C=OCH$_3$), 2.74 (t, J = 2.0 Hz, 1H, C≡CH), 3.38 (dd, J = 10.0 Hz, 8.0 Hz, 1H), 3.45–3.96 (m, 30H), 4.04 (d, J = 2.0 Hz, 2H, OCH$_2$C≡), 4.07 (d, J = 9.5 Hz, 1H), 4.10 (s, 2H, C=OCH$_2$O), 4.36 (d, J = 8.0 Hz, 1H, Fuc-H1), 4.99–5.08 (m, 2H, GlcNAc-H1 & Gal-H1), 6.98 (d, J = 9.0 Hz, 1H, ArCH)), 7.28 (d, J = 9.0 Hz, 2H, ArCH). **MS** (MALDI) m/z [M + Na]$^+$ Calculated = 915.358. Found = 915.521.

Synthesis of SiaLe^X-Ar-4-alkyne (74)

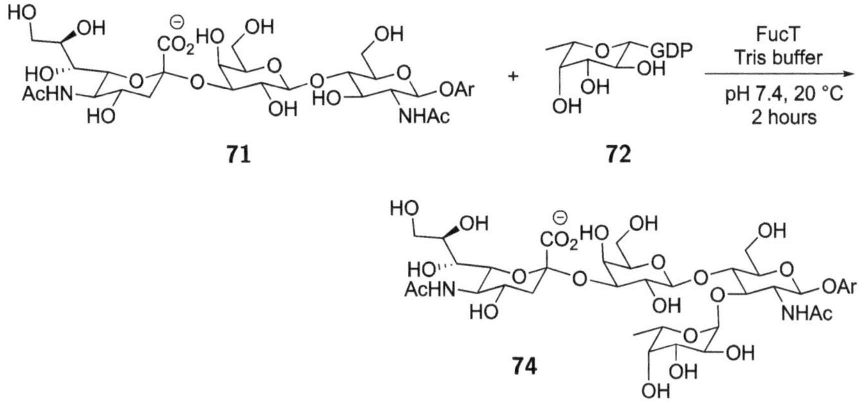

Adapted from a literature procedure [32]. Expressed and purified enzyme FucT (8 μL of solution at 6.3 mgmL^{-1}, so 50 μg) was added to Tris buffer (1 mL) at pH 7.4 with 20 mM $MnCl_2$, followed by BSA (1 mg), alkaline phosphatase (2 μL of 10 UμL^{-1} solution, so 20 U), and SiaLacNAc-Ar-4-alkyne (**71**, 4.5 mg, 4.3 μmol) and GDP-Fuc (**72**, 3.8 mg, 6 μmol). The reaction was left for 2 h at RT. After this time, precipitate was removed by centrifugation and the supernatant placed on a C18 Sep-Pak, which was washed with H_2O (5 mL) and then the products eluted with 3 mL each of 10%, 20%, 30%, 40%, and 50% MeOH in H_2O. The UV-active (by TLC) fraction—30%—was reduced under vacuum to give a viscous liquid. This was purified by semi-preparative HPLC (15-25% MeCN in H_2O) to give 2.2 mg of Le^X-Ar-4-alkyne (**73**, 1.9 μmol, 41% yield). Novel compound. ^{1}**H NMR** (400 MHz, D_2O) $\delta_H = 1.05$ (d, $J = 6.5$ Hz, 3H, Fuc-CH_3), 1.89 (s, 3H, C=OCH_3), 1.90 (s, 3H, C=OCH_3'), 1.93 (s, 1H, Sia-H3), 2.64 (dd, $J = 12.5$ Hz, 4.5 Hz, 1H, Sia-H3′), 2.74 (t, $J = 2.0$ Hz, 1H, C≡CH), 3.37–4.00 (m, 38H), 4.01–4.13 (m, 5H), 4.43 (d, $J = 8.0$ Hz, 1H, Fuc-H1), 4.99–5.06 (m, 2H, GlcNAc-H1 & Gal-H1), 6.98 (d, $J = 9.0$ Hz, 1H, ArCH)), 7.28 (d, $J = 9.0$ Hz, 2H, ArCH). **MS** (MALDI) m/z [M-H]$^-$ Calculated = 1182.457. Found = 1185.890.

7.9 Protein Expression

7.9.1 Expression of α-1,3-Fucosyltransferase from H. pylori

Acknowledgement: this work was carried out with Dr David Kwan (University of British Columbia, Vancouver, Canada).

Sequences for α-1,3-Fucosyltransferase (FucT) from *Helicobacter pylori* with *N*-terminally fused 6× His tag. Published protein sequence [32].

Amino acid sequence:

MFQPLLDAYVESASIEKMASKSPPPLKIAVANWWGDEEIKEFKNSVLYFILSQRYTITLHQNPNEFSDL

VFGNPLGSARKILSYQNAKRVFYTGENESPNFNLFDYAIGFDELDFNDRYLRMPLYYDRLHHKAESVND

TTAPYKLKDNSLYALKKPSHCFKEKHPNLCAVVNDESDPLKRGFASFVASNPNAPIRNAFYDALNSIEP

VTGGGSVRNTLGYNVKNKNEFLSQYKFNLCFENTQGYGYVTEKIIDAYFSHTIPIYWGSPSVAKDFNPK

SFVNVHDFKNFDEAIDYIKYLHTHKNAYLDMLYENPLNTLDGKAYFYQNLSFKKILAFFKTILENDTIY

HDNPFIFCRDLNEPLVTIDDLRVNYDDLRVNYDDLRINYDDLRVNYDDLRINYDDLRVNYDDLRVNYLE

HHHHHH

Calculated average isotopic mass = 49 229

A_{280} = 64 430

Escherichia coli BL21 (DE3) transformed with a purchased plasmid with the sequence for the expression of FucT was grown in a 10 mL starter culture of lysogeny broth (LB) supplemented with 100 mg mL^{-1} ampicillin overnight at 37 °C. The next day this culture was used to innoculate 500 mL further of supplemented LB, which was incubated at 37 °C until it reached A_{600} of 0.5 (3 h). Expression was then induced by the addition of IPTG (240 mg, 1 mmol) and the culture incubated at 25 °C for a further 5 h. The cells were collected by centrifugation (6000 rpm, 4 °C, 30 min), the supernatant discarded, and the cells stored at −20 °C overnight. The next day, the cells were resuspended in 40 mL of 'buffer A' (50 mm Tris, 500 mm NaCl in H_2O) supplemented with DNAse, protease inhibitor, and lysozyme. The cells were lysed using a homogeniser and the cell fragments removed by centrifugation (15 000 rpm, 4 °C, 30 min). The supernatant was then purified using an *Äkta* Pure FPLC equipped with a HiTrap column. A gradient of buffer A and buffer B (50 mM Tris, 50 mm NaCl, 50 mm imidazole in H_2O) (0–100% buffer B over 25 min at 1 $mLmin^1$). Fractions were analysed by sodium dodecyl sulfate polyacrylamide gel electrophoresis (SDS-PAGE) using a 4–12% PAGE gel:

Lanes:

1 = molecular weight marker, with masses marked (in kDa); 2 = insoluble fragments; 3 = supernatant; 4 = flow-through; 5–13 = odd fractions 5–21 of gradient.

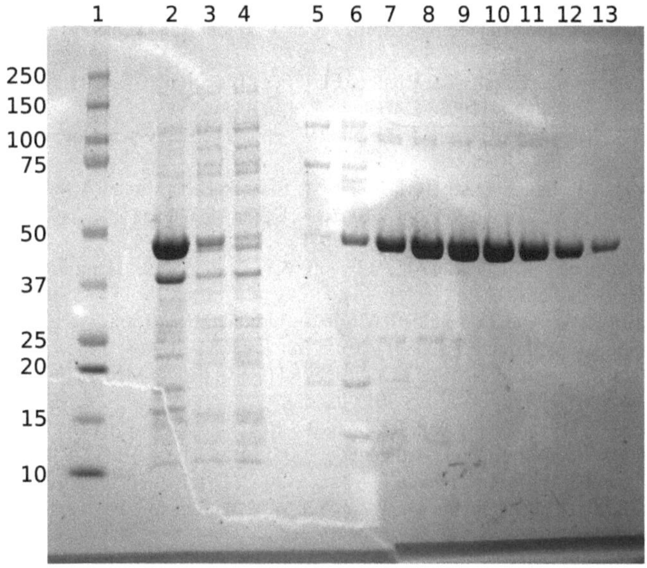

Fractions 9–21 of the gradient were collected and concentrated to 250 μL using a centrifugal concentrator (30 kDa cutoff), and the concentrated protein solution was diluted with 7.5 mL Tris buffer (pH 8.0). The concentration procedure was repeated, and dilution with 1 mL Tris buffer gave a protein solution at approximately 6.3 mg mL^{-1} (determined by A_{280}).

References

1. Rojas, M. GlyVicinity. http://glycosciences.de/tools/glyvicinity/
2. Lütteke T, Frank M, von der Lieth C-W (2005) Nucl Acids Res 33:D242–D246
3. Wang ZD, Matin M, Sheikh S (2005) Molecules 10:1325–1334
4. Dunbrack R PISCES: a protein sequence culling server. http://dunbrack.fccc.edu/PISCES.php
5. Li W, Godzik A (2006) Bioinformatics 22:1658–1659
6. CD-HIT. http://weizhongli-lab.org/cd-hit/
7. Agirre J, Iglesias-Fernández J, Rovira C, Davies GJ, Wilson KS, Cowtan KD (2015) Nat Struct Mol Biol 22:833–834
8. The PyMOL Molecular Graphics System, Version 1.7.0.3. (2014)
9. Brandl M, Weiss MS, Jabs A, Sühnel J, Hilgenfeld R (2001) J Mol Biol 307:357–377
10. Hudson KL, Bartlett GJ, Diehl RC, Agirre J, Gallagher T, Kiessling LL, Woolfson DN (2015) J Am Chem Soc 137:15152–15160
11. Mayato C, Dorta RL, Vázquez JT (2004) Tetrahedron: Asymmetry 15:2385–2397
12. Cheshev P, Marra A, Dondoni A (2006) Carbohydr Res 341:2714–2716
13. Demeler B, Royal society of chemistry. In Scott DJ, Harding S, Rowe A (eds) Modern analytical ultracentrifugation: techniques and methods, pp 210–229
14. Hurton T, Wright A, Deubler G, Bashir B (2012) Sedimentation intrepretation program. http://sednterp.unh.edu/
15. Allerton R, Fletcher HG Jr (1954) J Am Chem Soc 76:1757–1760
16. Hernandez E, Galan A, Rovira C, Veciana J (1992) Synthesis 11:1164–1169

17. Diot J, García-Moreno MI, Gouin SG, Ortiz Mellet C, Haupt K, Kovensky J (2009) Org Biomol Chem 7:357–363
18. Sun X-L, Stabler CL, Cazalis CS, Chaikof EL (2006) Bioconjug Chem 17:52–57
19. Mereyala HB, Gurrala SR (1998) Carbohydr Res 307:351–354
20. Fernandes A, Viterisi A, Aucagne V, Leigh DA, Papot S (2012) Chem Commun (Cambridge, U.K.) 48, 2083–2085
21. Nierengarten J-F, Iehl J, Oerthel V, Holler M, Illescas BM, Muñoz A, Martín N, Rojo J, Sánchez-Navarro M, Cecioni S, Vidal S, Buffet K, Durka M, Vincent SP (2010) Chem Commun (Cambridge, U.K.) 46, 3860–3862
22. Rowan AS, Nicely NI, Cochrane N, Wlassoff WA, Claiborne A, Hamilton CJ (2009) Org Biomol Chem 7:4029–4036
23. Nakabayashi S, Warren CD, Jeanloz RW (1986) Carbohydr Res 150:C7–C10
24. Jung JH, Shinkai S, Shimizu T (2002) Chem - Eur J 8:2684–2690
25. de Barros ALB, Cardoso VN, Mota LDG, Alves RJ (2010) Bioorg Med Chem Lett 20:315–317
26. Holaday JW (2011) Magistro, P. US 2011/0245287
27. Zhu B, Zhao C, Young FI, Franklin RJ, Song B (2014) Curr Protoc Stem Cell Biol
28. Kim Y-W, Lee SS, Warren RAJ, Withers SG (2004) J Biol Chem 279:42787–42793
29. Su DM, Eguchi H, Yi W, Li L, Wang PG, Xia C (2008) Org Lett 10:1009–1012
30. Persson K, Ly HD, Dieckelmann M, Wakarchuk WW, Withers SG, Strynadka NC (2001) Nat Struct Biol 8:166–175
31. Sugiarto G, Lau K, Qu J, Li Y, Lim S, Mu S, Ames JB, Fishe AJ, Chen X (2012) ACS Chem Biol 7:1232–1240
32. Lin S-W, Yuan T-M, Li J-R, Lin C-H (2006) Biochemistry 45:8108–8116

MIX
Papier aus verantwortungsvollen Quellen
Paper from responsible sources
FSC® C105338

If you have any concerns about our products,
you can contact us on
ProductSafety@springernature.com

In case Publisher is established outside the EU,
the EU authorized representative is:
Springer Nature Customer Service Center GmbH
Europaplatz 3, 69115 Heidelberg, Germany

Printed by Libri Plureos GmbH
in Hamburg, Germany